From Ada

The Numbers Game

From Adam to Noah—The Numbers Game

Why the Genealogy Puzzles of Genesis 5 and 11 Are in the Bible

Leonard Timmons

00198

SLIDING STORIES

SLIDING STORIES, LLC

Copyright © 2012 by Leonard Timmons

ALL RIGHTS RESERVED. Permission must be obtained to reproduce this book or any portion of this book in any form whatsoever except as provided by the Copyright Law of the United States of America. To obtain permission, please contact Sliding Stories. Contact information can be found at slidingstories.com.

Published by Sliding Stories, LLC | Duluth, GA

This book contains material reproduced from the following sources:

Tanakh: The Holy Scriptures: The New JPS Translation to the Traditional Hebrew Text, © 1985 by The Jewish Publication Society, with the permission of the publisher. (Referenced in the text as JPS Tanakh.)

Revised Standard Version of the Bible, copyright 1952 [2nd edition, 1971] by the Division of Christian Education of the National Council of the Churches of Christ in the United States of America. Used by permission. All rights reserved. (Referenced in the text as RSV.)

The Holy Bible: Containing the Old and New Testaments, © 1901 [Standard Edition] by Thomas Nelson & Sons. (Referenced in the text as ASV.)

The American Heritage Dictionary of the English Language, Fourth Edition, © 2009 by Houghton Mifflin Harcourt Publishing Company. Reproduced by permission.

And I would like to express my considerable gratitude to the copyright holders.

ISBN-13: 978-0-9833831-0-9 (Paperback)
ISBN-13: 978-0-9833831-1-6 (Hardback)
ISBN-13: 978-0-9833831-2-3 (eBook)
Library of Congress Control Number: 2011903237

Cover & Interior Design by Daniel Middleton | Scribe Freelance Book Design Company

Printed in the United States of America
17 16 15 14 13 12 11   2 3 4 5 6 7

Publisher's Cataloging-in-Publication

Timmons, Leonard.
From Adam to Noah--the numbers game : why the
genealogy puzzles of Genesis 5 and 11 are in the Bible /
Leonard Timmons. -- 1st ed.
p. cm.
Includes bibliographical references and index.
LCCN 2011903237
ISBN-13: 978-0-9833831-0-9 (pbk.)
ISBN-10: 0-9833831-0-3 (pbk.)
ISBN-13: 978-0-9833831-1-6 (hardback)
ISBN-10: 0-9833831-1-1 (hardback)
[etc.]
1. Bible O.T. Genesis V--Criticism, interpretation, etc. 2. Bible O.T. Genesis XI--Criticism,
interpretation, etc. 3. Numbers in the Bible. 4. Genealogy in the Bible.   I. Title.

BS1238.N86T56 2011   222'.1106
QBI11-600041

This book printed on acid-free paper.

*To those ancient teachers
who helped those with insight
gain more insight*

May their successors flourish

Contents

Illustrations

TABLE

Preface & Acknowledgments

I'VE SUSPECTED AS MUCH SINCE I was in high school. You might have been suspicious as well. Was something hidden in the genealogy of Adam in Genesis Chapter 5? How could Adam live for 930 years? How could Methuselah live for 969 years? I didn't know what the numbers meant and the explanations were unsatisfying. I did know that thousands of years ago, someone thought it was important to tell a story about people living to extreme ages. Now I know—I have discovered—why those ancient storytellers composed a story about a man who lived for 930 years and a man who lived for 969 years. The story is an important story, and the ages are important numbers.

You can find lots of theories about the meaning of the long lifetimes in Genesis 5. They range from theories at the lunatic fringe, to theories that are harmlessly crazy, to theories that are almost reasonable.[aa, ab, ac] If you've investigated these theories, you know that most of them start out crazy and many end up at the lunatic fringe. So your experience would tell you that my "discovery" is probably one more crazy theory from someone who doesn't really know what the genealogy means. Let me address your concerns. Even though this book is nonfiction, *knowing too much about the contents ahead of time can spoil the experience of reading it. So this is your spoiler alert.* But if you have to know that reading this will not be a waste of your time, let me refer you directly to Figure 2.5 and Figure 2.8. Look at them now if you'd like. Even if you don't fully understand the diagrams, you will be able to see at a glance that this analysis of the ages of the patriarchs is not a theory from the fringe. It is real. You may think I'm crazy by the time you finish reading this, but after you're done reading this book, you will *know* what the numbers in Genesis 5 really mean.

The genealogy is a calendar puzzle.

But the puzzle is bigger than Genesis 5. Genesis 11 also contains a numerical genealogy whose function is to confirm that the solution to the Genesis 5 puzzle is, in fact, the correct one.

But what we're encountering here is something much larger than the puzzles in Genesis 5 and 11. The whole of Genesis 1 to 11 is a collection of stories woven into a single fabric. The stories of Creation, the Garden of Eden, Abel and Cain, and the Tower of Babel are all a part of this larger story. Since the genealogy is not a real genealogy, you might also expect that the stories associated with it are not what they seem to be. Well, you'd be right. They aren't. These stories are also puzzles, and it will become clear that the authors intended those puzzles to be solved and interpreted. The calendar puzzle allows us to look at these stories in this new context. And given this new context, those interpretations will not be like any you may have encountered before. I expect that many readers will disagree with them. But I can tell you that I discovered the calendar puzzle using the very same techniques I was using when I first recognized that some of these stories were not what they seemed to be.

What I've discovered here is much larger than Genesis 1 to 11. Why should a calendar puzzle exist within the Bible at all? What was the Bible to those who composed its texts 3,500 years ago? What is it to us? The answers to these questions must change now that we know that the ages in Genesis 5 are a calendar puzzle. So, in order to understand those answers, I will introduce a rationale for the existence of the culture that produced these puzzles and what I believe to be the philosophy held by those who wrote the Bible. I will also examine whether that philosophy has meaning for us today.

Let me say that I am not a biblical scholar, so this cannot be a work of biblical scholarship. I will include references to books and websites that are accessible, clearly written, and have permanent links. (When I include a reference to a book or a website, I'm not endorsing it or the views held by its authors. You should also keep in mind that website content can change and may have changed since I accessed it.)

Scholars known as biblical critics have a theory about who wrote the first five books of the Bible, and I am familiar with that theory. Those scholars research the issues concerning when these books were first written and subsequently edited.[ad] But for most people, Moses is still considered the author of these books. Any other view would be quite controversial. I certainly cannot say who wrote the first five books of the Bible with any kind of authority. I can say, however, that solving the calendar puzzle of Genesis 5 adds to biblical scholarship, and it is my hope that this book will help everyone understand what they are reading when they open the Bible to read it.

Let me also say that I am not a calendar scholar and especially not an expert on calendars used in ancient Israel or elsewhere in that region. I have read good books on these subjects, however. Chief among those is *The Genesis Calendar: The Synchronistic Tradition in*

Genesis 1–11 by Bruce K. Gardner, which develops a theory of what the genealogy of Genesis 11 really means. The Reverend Dr. Gardner proposed a theory on the meaning of the numbers in Genesis 11 and followed that theory to its logical conclusion. His book clearly sets forth the requirements that anyone who would claim to have solved the genealogy puzzles must satisfy. Those requirements are satisfied here.

In addition, *Marking Time: The Epic Quest to Invent the Perfect Calendar* by Duncan Steel has been an invaluable reference for me as I attempted to put my rough understanding of the calendar on more solid ground.

Almost everyone knows about the genealogy in Genesis 5. It flatly states as fact something that is extremely hard for reasonable people to believe. If the storytellers were trying to call attention to the story, they certainly got mine. I've been focused on the genealogy of Adam for much of my life. I've come back to it again and again over more than 30 years. As I look back, I see a long journey of discovery, a journey no doubt traveled by many others over the thousands of years since these stories were first written down. At this journey's end, I can now say with certainty that something is hidden within these genealogies. This book is the product of that journey, a trek filled with lots of hard work and a good bit of fun.

I WOULD LIKE TO THANK one of my sisters, who gave me a large, four-function solar calculator for Christmas in 1999, on which I made the most important of my early discoveries. Who knew I needed a calculator?

I would also like to thank Ken Guyton, who has repeatedly read my work on this subject at its various stages of development and told me early on that I would need pictures—lots of them. Ken has made other valuable suggestions over that time. I would also like to thank Ken for being such a very good friend for the last 37 years. Let me also thank Ken's wife, Janie, for being such a wonderful wife to Ken and a wonderful mother to their children. Thanks for everything, Janie.

It gives me great pleasure to publicly thank Father Tom Francis of the Monastery of the Holy Spirit in Conyers, Georgia, for indulging me by looking at my early work when I was there on a week-long retreat. I do need to say that all that existed of this work were the early diagrams I had developed. I showed him everything up to the discovery of the importance of the number 416. I really appreciated his encouragement and excitement, since I was only looking for a "right track/wrong track" opinion from a dispassionate third party. I specifically wish to state that Fr. Francis does not know of, and therefore has expressed no opinion on, my biblical interpretation and should not be associated with those interpretations in any way. Even though I am not Catholic, I do appreciate the existence of the Monastery of the Holy Spirit and would like to encourage anyone who reads this to contribute to their continuing effort to know the mind of God. Please visit them in Conyers, Georgia, or at www.trappist.net.

I would also like to thank the following individuals who reviewed my manuscript: Dr. Jeremy Northcote (*The Paranormal and the Politics of Truth: A Sociological Account*); Dr. Philippe Guillaume (*Land and Calendar: The Priestly Document from Genesis 1 to Joshua 18*, *Waiting for Josiah: The Judges*, and *The Bible in Its Context*); Carol A. Hill (Making Sense of the Numbers of Genesis, *Perspectives on Science and Christian Faith*, vol. 55, no. 4, December 2003); Gary L. Havens (*Mathematical Design and Purpose of the Bible*, unpublished); Matthew Murry; and Barry Schieb.

I would also like to thank Jean Reber for her insightful criticism, and my editors, Peggy Emard and Elizabeth Polen, for their excellent work. They all helped me look at this book through their eyes.

This is also a great opportunity to publicly thank my mother, who under difficult conditions put forth the Herculean effort to care for all eleven of us and give us a great start in life under difficult conditions. She has always looked out for my best interests, and I love her dearly. I would also like to remember and thank my grandmother, Mrs. Julia Timmons. As a child I asked her questions she could not answer. She invited me to include her when I found the answers to those questions. The idea that it was possible for me as a child to know more than an adult was a *revelation*, and that realization defines my life.

Finally, my wife, Jeannie, has shown me that empathizing with and loving others is the best way to love yourself. I would like to recognize her contribution to this work and to my life—a life she helped save. Jeannie (a.k.a. Shirley) has offered me her considered opinion, for which I am very grateful.

The figures in this book are available as color posters and fliers.

Please visit

www.AdamToNoah.com

to obtain copies for your organization.

CHAPTER 1

The Genealogy of Genealogies

1.1 Three Questions

I'VE DISCOVERED THAT there is a highly accurate calendar hidden within chapter 5 of Genesis in the Bible. It's amazing. It's surprising. It's unbelievable.

You're suspicious, and you should be. If you're a religious reader, you may be suspicious because all kinds of people make all kinds of claims about the book that is the foundation of your religious life. Most of those claims can be safely ignored. Your understanding of the Bible is rooted in religious faith, and any claim I could make about the book would not change your basic understanding of it. If you're an anti-religious reader, you are probably suspicious because you consider the Bible to be a collection of ancient superstitions and myths that people misuse out of ignorance. From the anti-religious perspective, no claim I could make about the Bible would impact your basic understanding of it. Nothing I could say would make the Bible more relevant to you or your life.

If there is a calendar hidden in Genesis 5 that is as accurate as our modern calendar, that fact must change the way you look at the Bible. That change could be large enough to have an impact on your most fundamental understanding of the book. So I'm going to show you the calendar as quickly as I can. Since you should not believe my claim, I will try to prove that the calendar is actually there. If I can prove it to your satisfaction, then you'll want to know the answer to the following questions:

1) How is it that you were the one to make this discovery if the ages in Genesis 5 have been in plain sight for 2,500 to 3,500 years?

2) Why would the people who wrote the Bible hide a calendar in their sacred writings, and why would they hide it in this fashion?

3) If the Bible hides a calendar in one of its defining stories, then what *is* the Bible?

After I show you the calendar, I will begin the task of answering these three questions. I hope the answers will help you change your understanding of Genesis, of the Bible, and of the people who wrote it.

Let me warn you that what you're about to read is unconventional. If you are a highly religious person and the Bible is an object of devotion for you, then the answers to the questions above might not be something you want to know. If you are an anti-religious person and the Bible has been a target of your anger or ridicule, then you may also not want to know the answers to these questions. Some of you view the world from a religious perspective, while others of you see it from an anti-religious one. These two views of the world have been at odds for a very long time. I intend to merge these theistic and atheistic views into a single concept that I think flows naturally from the Bible writers. Merging the two concepts may be unpleasant and hard to understand. Let me encourage you to try. I will not ask you to change your beliefs or behavior. I will present a view of the world that I think the people who wrote the Bible held. I will ask that you see the world through their eyes. You would have to accept the notion that the Bible writers held the views that I attribute to them, of course. So if you cannot accept that I have my finger on the pulse of this ancient culture, that's OK. I am, however, earnestly trying to answer these three questions.

As a reader of any analysis of the Bible, and as a reader of any historical analysis, you should be keenly aware that each person approaches an analysis with his or her own biases. This is especially true of biblical analyses. People feel so strongly about the Bible and its contents that they *cannot* look at it dispassionately. Many have based their entire lives on the book. I grew up in a religious Christian culture, so you should take that into account while reading this. I have tried to be as unbiased and dispassionate as possible. But you should know that my understanding of the Bible and its authors is significantly different from that of the vast majority of people.

I think of myself as an amateur scientist, and I really wanted to present this material in that kind of rigorous scientific style. Unfortunately, very few people would care to read such a book. So I determined that I would use a "textbook lite" style. I've included numbered chapters and sections so that those who may want to discuss this book with others can precisely refer to the sections under discussion. A step-by-step proof that a calendar can be found in

Genesis 5 might be a bit difficult to follow. And I really do want you to understand what I'm saying, so I've tried to show you that proof clearly and with a minimum of jargon. I've tried to do the same with the remainder of the book as well.

In real life I am an engineer, and it is my job to make things work, not to generate a proof that they *can* work. I have written this book in that engineering spirit, so making everything work together is my ultimate goal. If we look back through history, we find that scientists as we know them today simply did not exist in ancient times. Ancient scientists were all engineers—they had to make things work. Even ancient philosophers were engineers—their systems of thought had to achieve practical goals. So if my answers don't work for you, then you have my sincere apologies. I have tried to make all of this work together, though I could be wrong about it all. I don't think so, of course.

One thing that might not be clear at this point is that a book about a calendar hidden within the text of the Bible will use a lot of numbers. Calendars are numerical documents, and we will manipulate a lot of numbers to extract that calendar. So I've tried to reduce the use of numbers through the use of figures. I've tried to provide enough figures to allow you to understand the basics just by looking at them. I want you to feel free to look at the figures and skim the text in the really number-heavy parts of the book. I hope this helps those of you who have problems with numbers.

1.2 Genesis 5

When I was a child, I was taught that the oldest man to ever live was Methuselah, who lived to be 969 years old. It was a great number to remember, as I recall. Eventually I read Genesis 5 with all of its "begats." We were studying the King James Version, and when that version was published in 1611 CE[1], a father "begat" his son. Even though I found the KJV hard to understand, it was easy to notice that most of the people listed in Genesis 5 lived for more than 900 years. My Sunday School teacher told us that the story was true. She said that people lived to extraordinary ages in ancient times, but as time passed our ancestors did things that were more and more evil. The corruption that resulted caused people to die sooner than they

---

1   The Common Era. CE is equivalent to AD.

otherwise would have. Because everyone had corrupted themselves, God set an upper limit on our lifetimes to 120 years.

In middle school, I began to question the story. Could the writers have meant something else? It seemed to be a physical impossibility for anyone to have ever lived into their 900th year. I first thought to scale the ages into a range we would recognize as possible. The idea behind scaling is that the ages are not given in real years, but in months or some other unit that the Bible writers call "years." So Methuselah's age of 969 "years" could be 969 months, or about 80 years. But ultimately scaling didn't work. If I scaled the lifetimes so that their values were reasonable, some fathers would have had sons at unreasonably young ages. I didn't see a good way to resolve this dilemma, so I put the problem away. But I was drawn back to it again and again.

As I continued my reading and research of the Bible, I began to realize the importance the Bible writers placed on actually understanding things. Once I noticed this emphasis, I counted the number of times Jesus is reported to have used the word "understanding" during his ministry. I soon realized how little I knew about what was really in the Bible. So I read the entire thing a couple of times. I tried a parallel Bible and ultimately bought a copy of the Tanakh, published by the Jewish Publication Society. I liked the JPS Tanakh because the translators seemed to make a substantial effort to translate without imposing their own biases.

As a Christian, I spent a lot of time in the New Testament studying the life and the words of Jesus. His parables amazed me. It was so interesting and ultimately surprising to see Jesus explain his mysterious parables to his disciples. The surprising part was that his explanations were parables themselves. Explaining a parable with a parable produced no net explanation! Why would anyone do that? As the story continued, Jesus told his disciples that he spoke to the assembled masses in parables so that they would *not* be able to understand him. That was a revelation. And in the book of Revelation, I had some success interpreting Chapter 13. I slowly began to realize that the Bible was filled with riddles—really hard riddles. And as a book, it was intended to confuse, not to explain. I could see that clearly, but I did not understand why this would be so.

I returned to Genesis with this idea and had some success interpreting the story of Noah's ark and the story of the Garden of Eden.[ae] Once I had a handle on those, I returned to the genealogy of

Genesis 5 and spent a lot of time and effort trying to figure it out. I tried the scaling idea again and eventually dismissed it altogether. I tried lots of hypotheses that did not work out, more than I can possibly remember.

Eventually I turned down a path I had known about but never recognized as significant. I knew that the genealogy in Genesis 5 was not just the genealogy of Adam and his son Seth but was closely related to the genealogy of Cain in the preceding chapter. The thing that stood out to me, that caught my attention, was that the names of the patriarchs in the two genealogies were so similar. I explored the idea that the genealogy of Cain was the Bible writers' way of telling us how to solve the problem of the genealogy of Seth. And it was.

Here is the text of both genealogies:

> [Genesis 4:17–26]
> Cain knew[2] his wife, and she conceived and bore Enoch. And he then founded a city, and named the city after his son Enoch. To Enoch was born Irad, and Irad begot Mehujael, and Mehujael begot Methusael, and Methusael begot Lamech. Lamech took to himself two wives: the name of the one was Adah, and the name of the other was Zillah. Adah bore Jabal; he was the ancestor of those who dwell in tents and amidst herds. And the name of his brother was Jubal; he was the ancestor of all who play the lyre and the pipe. As for Zillah, she bore Tubal-cain, who forged all implements of copper and iron. And the sister of Tubal-cain was Naamah.
>
> And Lamech said to his wives,
>
> > "Adah and Zillah, hear my voice;
> > O wives of Lamech, give ear to my speech.
> > I have slain a man for wounding me,
> > And a lad for bruising me.
> > If Cain is avenged sevenfold,
> > Then Lamech seventy-sevenfold."
>
> Adam knew his wife again, and she bore a son and named him Seth, meaning, "God has provided me with another offspring

---

2   To "know" in this fashion means carnal knowledge, a euphemism for sexual intercourse.

in place of Abel," for Cain had killed him. And to Seth, in turn, a son was born, and he named him Enosh. It was then that men began to invoke the LORD by name.

[Genesis 5:1–32]
This is the record of Adam's line.—When God created man, He made him in the likeness of God; male and female He created them. And when they were created, He blessed them and called them Man.—When Adam had lived 130 years, he begot a son in his likeness after his image, and he named him Seth. After the birth of Seth, Adam lived 800 years and begot sons and daughters. All the days that Adam lived came to 930 years; then he died.

When Seth had lived 105 years, he begot Enosh. After the birth of Enosh, Seth lived 807 years and begot sons and daughters. All the days of Seth came to 912 years; then he died.

When Enosh had lived 90 years, he begot Kenan. After the birth of Kenan, Enosh lived 815 years and begot sons and daughters. All the days of Enosh came to 905 years; then he died.

When Kenan had lived 70 years, he begot Mahalalel. After the birth of Mahalalel, Kenan lived 840 years and begot sons and daughters. All the days of Kenan came to 910 years; then he died.

When Mahalalel had lived 65 years he begot Jared. After the birth of Jared, Mahalalel lived 830 years and begot sons and daughters. All the days of Mahalalel came to 895 years; then he died.

When Jared had lived 162 years, he begot Enoch. After the birth of Enoch, Jared lived 800 years and begot sons and daughters. All the days of Jared came to 962 years; then he died.

When Enoch had lived 65 years, he begot Methuselah. After the birth of Methuselah, Enoch walked with God 300 years;

and he begot sons and daughters. All the days of Enoch came to 365 years. Enoch walked with God; then he was no more, for God took him.

When Methuselah had lived 187 years, he begot Lamech. After the birth of Lamech, Methuselah lived 782 years and begot sons and daughters. All the days of Methuselah came to 969 years; then he died.

When Lamech had lived 182 years, he begot a son. And he named him Noah, saying, "This one will provide us relief from our work and from the toil of our hands, out of the very soil which the LORD placed under a curse." After the birth of Noah, Lamech lived 595 years and begot sons and daughters. All the days of Lamech came to 777 years; then he died.

When Noah had lived 500 years, Noah begot Shem, Ham, and Japheth.

[Following this genealogy, we read the story of Noah before the Flood. Then Genesis 7:6 tells us:]

Noah was 600 years old when the Flood came, waters upon the earth.

[The story of the Flood follows. At the end of the story, Genesis 9:28 says:]

Noah lived after the Flood 350 years. And all the days of Noah came to 950 years; then he died.[af]

As you can see from the text, the genealogy of Seth is interwoven with the story of Noah and the story of the Flood. I have removed the sections containing the story of the Flood for clarity and brevity. The text shows that both genealogies have an Enoch. Each also has a Lamech. Cain's genealogy has a Methusael, while Seth's genealogy has a Methuselah. And both Methusael and Methuselah give birth to a son named Lamech. While this might be an accident of history, I decided it was much more likely that the writers were trying to tell us something.

So what is the purpose of Cain's genealogy? His descendants were responsible for some early inventions, but those inventions were not central to the genealogy. This genealogy is simpler than that of Genesis 5; no numbers denote when people were born or when they died. Cain's genealogy is exceptional because the women are named, but that also is not the central point of the genealogy. We come to the central point when Cain's descendant Lamech makes an amazing claim. He tells his wives that he has killed a man who accidentally(?) wounded him, and he's killed a young man who may have just bumped into him. Lamech is telling his wives and us that he would kill absolutely anyone for just about any reason. In other words, Lamech was a degenerate murderer. This was the piece of information that allowed me to realize that not only was the genealogy of Genesis 5 a riddle, it was also a puzzle of some significant ingenuity.

What I discovered was that the ages of the patriarchs in Genesis 5 are not, in fact, human lifetimes. What they are is a puzzle, a numerical riddle. They are the numerical equivalent of a crossword puzzle. Clues are given, and confirmations become apparent when you get the answers right. The genealogy is unlike a crossword puzzle in that I did not know its subject matter before I tried to solve it. When I've solved crossword puzzles in the past, I've always known what the puzzle was about. And as a puzzle whose subject matter was unknown, the genealogy of Genesis 5 was *extremely* hard to solve. It is only in the process of solving this puzzle/riddle that we find out what it is about. And it is this subject matter, this calendar, that the puzzle makers were trying to communicate to us from the foundations of our history.

CHAPTER 2

The Puzzle Solved

2.1 Clues

A CLOSER READING TURNS UP clue after clue that the genealogy of Seth is a puzzle. The first clue is that a lot of the ages are "round" numbers. Adam lived 130 years, had a son, then lived another 800 years. Enosh lived 90 years, had a son, then lived another 815 years. The round numbers end in "0" or "5." If the genealogy was describing real ages, we would expect the numbers to be less "round." But we could also conclude that the people who created this puzzle were guessing at the ages of the patriarchs because historically accurate information was unavailable.[ag, 3]

The second clue is that some of the ages are clearly chosen. Enoch's lifetime of 365 years is most certainly chosen to equal the number of days in a year. Lamech's lifetime of 777 years is clearly chosen to be a "nice" number[4]. So some of the numbers are "nice" numbers, and some are "round" numbers. Had the ages been taken from historical information on the actual lifetimes of the patriarchs, we would expect the ages to be both less "round" and less "nice."

Many of the numbers are "odd," however. They seem to be neither "round" nor "nice." These "odd" numbers include 807, 912, 162, 782, and 182. But on closer inspection, you might notice that 182 is half of 365. And this fact makes 182 a "nice" number, since 365 is a "nice" number. It also gives us the suspicion that some of the other numbers may be related to the number of days in a year. We might also suspect that this puzzle could be calendar-related (though I must admit that I did not have that suspicion when I made my first discoveries). But in any event, we should try to find the relationships that make the other "odd" numbers "nice" numbers.

---

3   The belief that the ages are taken from historically accurate information was widespread among scholars, even into the last half of the 19th century. Many people still hold this belief.

4   A nice number is one that is special in some way. Even today we think of numbers that have special characteristics, or numbers that are related to things we think of as important, as "nice" numbers.

The third clue that this genealogy is a puzzle requires some calculation and a diagram. The lives of Enoch and Lamech both end before their fathers' lives do. Enoch disappears before his father dies, and Lamech dies before his father does. This fact sets them apart and makes their lifetimes of 365 years and 777 years even more significant. Their names were significant originally because of the correspondence between the Enoch and Lamech in Cain's genealogy and the Enoch and Lamech of Seth's genealogy. That correspondence weakly implied that we should pay more attention to them.

The fourth clue requires more calculation and another diagram. But before I introduce any diagrams, let me explain their format. The lifetimes of the patriarchs can be plotted on a time line. We can show the flow of time from the top to the bottom of a diagram. However, since the lifetimes overlap, a collection of downward-pointing arrows could be hard to understand. So to make the graphs more compact and understandable, the lifetime of each patriarch is shown as a downward-pointing arrow and an arrow pointing to the left or right. The line changes direction at the birth of the patriarch's inheriting son.[5] See Figure 2.1.

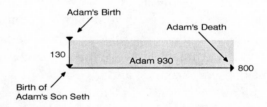

© 2011 Leonard Timmons

Figure 2.1: Adam's Lifetime. Adam lived 130 years before the birth of his son Seth. Then he lived another 800 years before he died. He lived a total of 930 years.

When we add a second patriarch to the time line, we point the arrow in the opposite direction to make the diagram more readable:

---

5   In this culture and at that time, the son who inherited is the first one mentioned in the genealogy. That son is not necessarily the firstborn son, as is clear in the case of Seth, the third son of Adam and Eve.

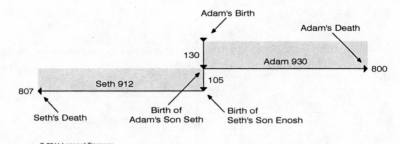

© 2011 Leonard Timmons

Figure 2.2: Adam's and Seth's Lifetimes. Adam lived 130 years before the birth of his son Seth. Then he lived another 800 years before he died. Adam lived a total of 930 years. Seth lived 105 years before the birth of his son Enosh. Then he lived another 807 years before he died. Seth lived a total of 912 years.

This format allows us to maintain a clear view of the passage of time, while keeping the time line compact and free of clutter. Figure 2.3 shows the full genealogy of Seth in this format:

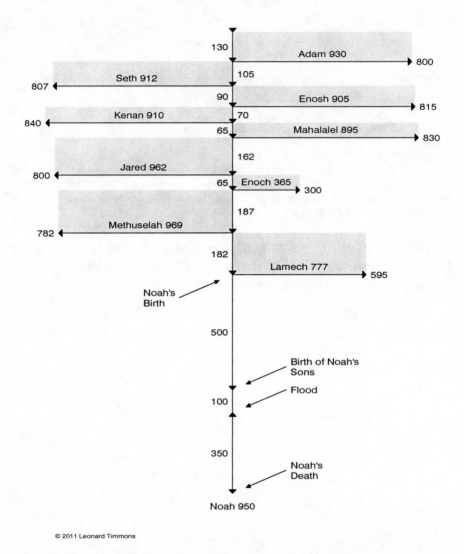

© 2011 Leonard Timmons

Figure 2.3: Genealogy of Seth. The lifetimes of the patriarchs of Genesis 5 are shown. The lifetime of Noah is shown without a right angle. The Flood is marked with an inverted arrowhead, since a number of events occur immediately before the Flood.

The fourth clue has to do with the hint given in the genealogy of Cain. Both genealogies have an Enoch and a Lamech. We have already noted that these two patriarchs are significant because each of them dies (or disappears) before his father dies. The critically important clue is that Cain's Lamech is a murderer. He will escalate and avenge any violence done to him 77 times, versus the 7 times God will escalate and avenge any violence done to Cain. That threat by Cain's Lamech contains a triple seven, and that 777 directly corresponds to the 777 years that the Lamech of Seth's genealogy lives. The use of three sevens in succession in both passages tells us clearly that there is a correspondence between the two. We get our first firm clue of what that something is when we note that seven patriarchs die during the lifetime of Seth's Lamech. See Figure 2.4:

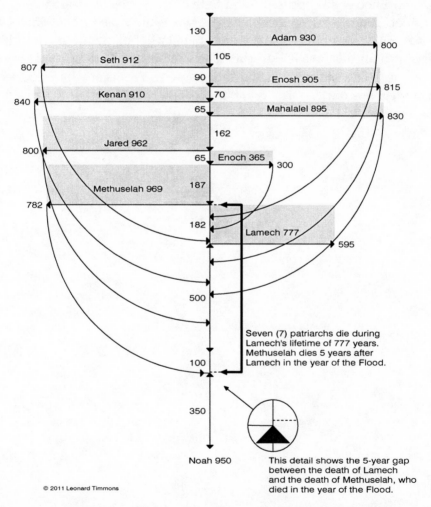

© 2011 Leonard Timmons

Figure 2.4: Lamech's Lifetime. During the lifetime of Lamech, seven patriarchs die. These deaths are alluded to in the genealogy of Cain, in which Cain's Lamech is a murderer.

Now we know that the direct correspondence between the genealogy of Cain and the genealogy of Seth is that Cain's Lamech was a murderer and seven patriarchs die during the lifetime of Seth's Lamech. The implication is clear, and this clue tells us that the *deaths* of the patriarchs are important in this puzzle—not their births.

2.2 The First Part: Adam, Enoch, And Seth

The first three patriarchs to die during the lifetime of Lamech are Adam, Enoch (disappeared), and Seth. Adam dies 56 years after the birth of Lamech. Enoch disappears 113 years after the birth of Lamech. And Seth dies 168 years after the birth of Lamech. Enoch's disappearance is $(2 \times 56) + 1$ years after the birth of Lamech, and Seth's death is 3×56 years after the birth of Lamech. This is an almost perfect pattern and is our confirmation that we have solved part of this puzzle. See Figure 2.5.

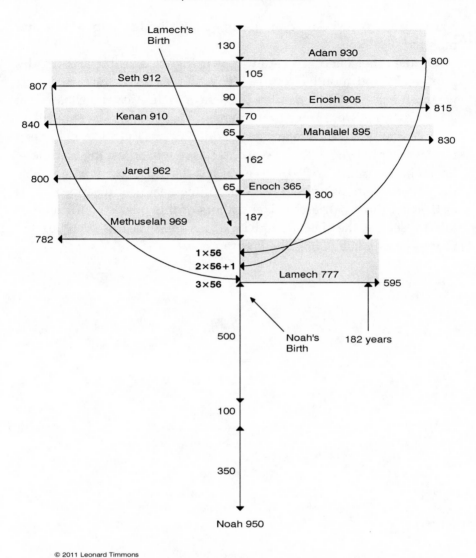

© 2011 Leonard Timmons

Figure 2.5: The First Three Patriarchs. After the birth of Lamech, the patriarchs die at nearly regular intervals. The intervals are multiples of 56 years. Adam dies at 1×56 years after the birth of Lamech. Enoch disappears at (2×56) + 1 years after the birth of Lamech. And Seth dies at 3×56 years after the birth of Lamech.

This puzzle handles the lifetime of Enoch in a special way. If Enoch had lived 364 years, then we would have a perfect 1×56, 2×56, 3×56 pattern for the first three deaths after the birth of Lamech. The authors hide the number 364 in this fashion to introduce us to their concept of the "Divisible Year." A 365-day year is not very divisible, although

365 is 5×73. A 365-day year is not easily divided into weeks and months.

On the other hand, a 364-day year is highly divisible and divides into weeks and months very easily. A 364-day year has exactly 52 weeks. It has exactly 13 months of 28 days each. A calendar based on a 364-day year would need an intercalary day[6] added at the end of each year and two intercalary days added every 4 years to help keep the calendar in sync with the seasons. So we will adjust the lifetime of Enoch in this puzzle to 364 + 1 years, and that adjustment will give us the perfect relationship that we seek for the deaths or disappearances of the first three patriarchs. The authors of this puzzle confirm all of this by setting the time period in which these deaths occur to 182 years—exactly half of 364. See Figure 2.6.

---

6   A day added to the year to synchronize the year with the seasons or some other astronomical event. Those of us who use the Gregorian calendar have our own intercalary day: February 29, or Leap Day, which is added to the calendar every four years. "calary" implies calendar.

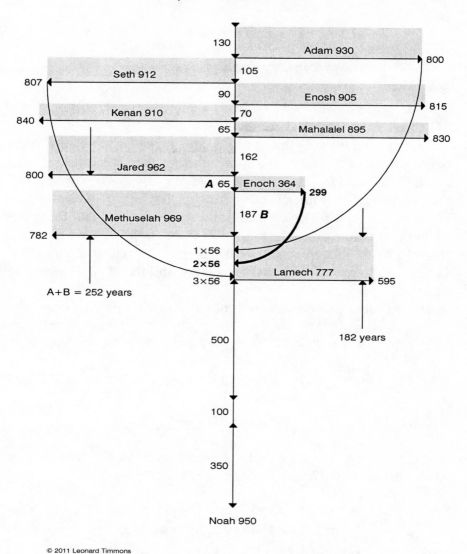

© 2011 Leonard Timmons

Figure 2.6: Enoch's Lifetime Adjusted. If we adjust the lifetime of Enoch to 364 years to equal the number of days in a "Divisible Year," then the first three patriarchs die at regular intervals of 56 years after the birth of Lamech.

The question that comes to mind here is why the authors of this puzzle would choose the number 56 as the basis of this portion of the puzzle. They chose it because the number 56 is (7×7) + 7, which gives us the triple seven of Lamech's lifetime again. We also need to look at the Divisible Year a little more deeply. The Divisible Year can be written as a sum of sevens in the following way:

$$77 + 77 + 77 + 77 + 7 + 7 + 7 + 7 + 7 + 7 + 7 + 7 = 364$$

or

$$77 + 77 + 77 + 77 + (7 \times 7) + 7 = 364$$

or

$$77 + 77 + 77 + 77 + 56 = 364$$

So the authors of this puzzle are repeating the idea that combinations of the number seven, whether they are concatenated, multiplied, or added, are "nice" numbers. We saw this behavior in the clues section where Cain's Lamech insisted that he would be avenged 77 times versus Cain's seven times, and we determined that the triple seven there pointed directly to the 777 years in the lifetime of Seth's Lamech. With this information it becomes crystal clear that the authors are "playing" with these numbers, and that this genealogy is a very sophisticated numbers game.

The Divisible Year can also be broken up in the following way:

$$(77 + 7 + 7) + (77 + 7 + 7) + (77 + 7 + 7) + (77 + 7 + 7) = 364$$

or

$$(84 + 7) + (84 + 7) + (84 + 7) + (84 + 7) = 364$$

or

$$4 \times 91 = 2 \times 182 = 364$$

We can merge these two sets of equations by noting that we can break up one of the 77s into the sum of 56 and 21:

$$77 + 77 + 77 + \quad (77) \quad + 56 = 364$$
$$77 + 77 + 77 + (21 + 56) + 56 = 364$$
$$(77 + 77 + 77) + (7 + 7 + 7) + 56 + 56 = 364$$

Then we can distribute the three sevens in the 21 to the three 77s that are left:

$$(77 + 7) + (77 + 7) + (77 + 7) + 56 + 56 = 364$$
$$84 + 84 + 84 + (2 \times 56) = 364$$

and we know that 364 is 4×91, so we get:

$$2 \times 56 + 3 \times 84 = 4 \times 91$$

All of these numbers are "nice" numbers because they are related to the Divisible Year. The authors chose the "nice" number 56 to separate the deaths of the first three patriarchs and to fit within a 182-year span. The authors chose the "nice" number 364 as the lifetime of Enoch and added one to make it a "nice," "round" number because the number of days in a year is a nice, round number by definition.

But what about the remaining numbers? The lifetime of Seth and the lifetime of Adam have to be set by the puzzle authors in some manner. You can see how they were set by studying Figure 2.7.

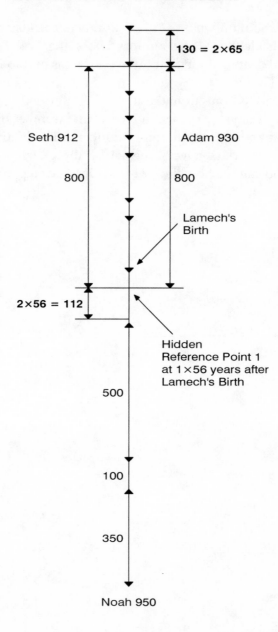

130 = 2×65

Seth 912   Adam 930

800   800

Lamech's Birth

2×56 = 112

Hidden Reference Point 1 at 1×56 years after Lamech's Birth

500

100

350

Noah 950

© 2011 Leonard Timmons

Figure 2.7: Lifetimes of Adam and Seth. At 56 years after the birth of Lamech (at 1×56 above), the authors of this puzzle created a hidden reference point from which they set the lifetimes of both Adam and Seth. The puzzle authors used an 800-year timespan bracketed by a 2×56-year span and a 2×65-year span. The number 65 was chosen because it is the reverse of 56 and hides information about the puzzle.

As you can see from Figure 2.7, the puzzle authors chose an 800-year span bracketed by a 2×56-year span and a 2×65-year span. The 800-year span was chosen because it is a nice, round number. And the 65 is used because it is the reverse of 56 and *seems* to be related to the 365 days in a year. Choosing both 800 and 65 somewhat arbitrarily makes this puzzle much harder to figure out. So Adam's lifetime of 930 years consists of an 800-year span and a 130-year span. This division is clear from the text. However, Seth's lifetime of 912 years consists of an 800-year span and a 112-year span. And that's not something you would know, even if you've made the Bible your life's work.

The puzzle authors set the 56th year of Lamech's lifetime as a reference point for the lifetimes of both Adam and Seth. It is at this point that the shared 800-year span in the ages of Adam and Seth ends. You can see this reference point at Hidden Reference Point 1 in Figure 2.7 (the "1×56" point). The *hidden reference point* makes this puzzle much more difficult to figure out.

The 112th year of Lamech's lifetime is a reference point for Enoch's lifetime. This is at the "2×56" point of Figure 2.6. Since Enoch's lifetime is set to 364 years before this point and one year after, the time of his birth is fixed relative to this reference point. As a consequence, there is a $364 - (2 \times 56) = 364 - 112 = 252$-year span between the birth of Enoch and the birth of his grandson Lamech ($A + B$ in Figure 2.6). This period is divided in two by the birth of Enoch's son Methuselah. His birth divides this 252-year period into a 65-year period (A in Figure 2.6) and a 187-year period (B in Figure 2.6). The 65-year period is chosen fairly arbitrarily. The remaining period of $187 = 182 + 5$ years contains yet another secret which will be revealed below.

At this point as the puzzle authors were building this puzzle, they had fixed everything for Adam, Seth and Enoch. We know their births, deaths, and the births of their first sons (except for Seth's first son, Enosh, who introduces the next part of the puzzle). In addition to these three patriarchs, the authors have also fixed a 252-year period in which they have set the birth of Enoch, the birth of his son Methuselah, and the birth of his son Lamech. The 252-year period ends at Lamech's birth, an important reference point for the entire puzzle.

The first portion of Methuselah's lifetime has also been fixed, since we know when his son Lamech is born. The remainder of

Methuselah's lifetime is not handled in this part of the puzzle. We do know that the remainder of his life was set to 782 years. We'll find out why below.

2.3 The Second Part of The Puzzle

The second part of this puzzle is solved in a manner similar to what we've just done above. The solution to this part is much more complex, however. To explain and justify the solution, I will present the entire solution and then explain it in sections. The full solution is diagrammed in Figure 2.8:

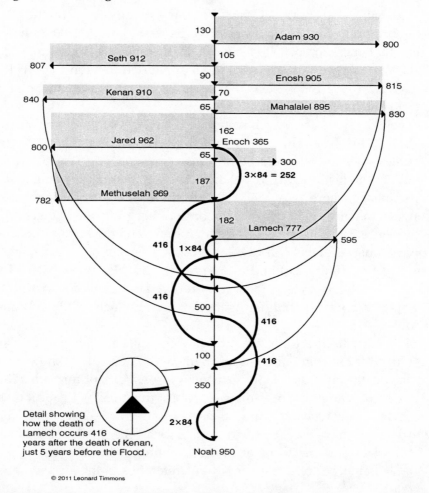

© 2011 Leonard Timmons

Figure 2.8: The Numbers Game. The full solution to the second part of this puzzle is quite complex. This portion does not include Methuselah, however. Nor does it include Adam, Enoch, or Seth.

This is a complex diagram, so let's take it apart piece by piece. (It was at this point that I began to think that this puzzle might contain a calendar.) I first found the pattern of 416-year intervals between the deaths of the remaining patriarchs. From this I knew that the 416-year interval was a part of the solution. The most obvious 416-year interval is defined by the death of Lamech and the death of Kenan. The other obvious 416-year period is defined by the birth of Lamech and the death of Mahalalel. The third 416-year interval spans the time between the death of Enosh and the birth of Noah's sons. The fourth 416-year interval that begins at the death of Jared is not explicitly marked at its other end. Its implicit ending is provided as a confirmation that we have actually solved this part of the puzzle. More on this later.

In the previous section we found a 252-year span [364 − (2×56)] between the birth of Enoch and the birth of Lamech ($A + B$ in Figure 2.6 and explicitly marked there). This span can be divided into three sections of 84 years each. The number 84 in this part of the puzzle performs the same function as the number 56 did in the first part. In that part the number 56 was symbolic of the triple seven of 777, because it is (7×7) + 7. In this part of the puzzle the number 84 symbolizes the triple seven of 777 because 84 is 77 + 7. The number 84 in this part of the puzzle is the logical successor to the number 56 in the first part of the puzzle. Just as the number 56 was used to confirm the correct solution in the first part of the puzzle, the number 84 is used for confirmation in this part.

As you can see from Figure 2.8, a 1×84, 2×84, and 3×84 sequence is provided as confirmation that this part of the puzzle has been solved. As we discussed above, a 3×84-year span separates the birth of Enoch and the birth of Lamech and is connected to the 416-year span from the birth of Lamech to the death of Mahalalel. Next, the death of Enosh separates a 1×84-year span from a 416-year span and these remarkably sum to the 500 years of Noah's life before the birth of his sons. Finally, the 2×84-year span that is connected to the 416-year span after the death of Jared marks the death of Noah at the very end of Seth's genealogy. The fact that the junction between the 416-year span that follows the death of Jared and the 2×84-year span that comes after it is not marked is an additional confirmation that we have solved this part of the puzzle.

The number 416 is very important in this part of the puzzle. Is there anything significant about it other than that it is 500 − 84? Well, it also happens to be 364 + 52, which is the number of days in a "Divisible Year" and the number of weeks in that year. More on this later.

Note again that the death of Methuselah, denoted by the upturned arrow (see the insert in Figure 2.8) is not included in this part of the puzzle. Methuselah dies in the 600th year of Noah's life, when the Flood occurs. Lamech dies in the 595th year of Noah's life, five years before the Flood. So Methuselah's life is the third part of this three-part puzzle. We will examine Methuselah's lifetime below.

2.3.1 Enosh (Not Enoch)

When the death of Enosh (not Enoch) is fixed at 84 years after the birth of Noah and 416 years before the birth of Noah's sons, the importance of the 84 + 52 + 364 = 500 relation is emphasized. The puzzle authors also set Enosh's birth 105 years after the birth of his father, Seth. The rationale they used to fix the time of his birth, to fix the length of his lifetime, and to place the (1×84) + 416 = 500-year period is shown in Figure 2.9.

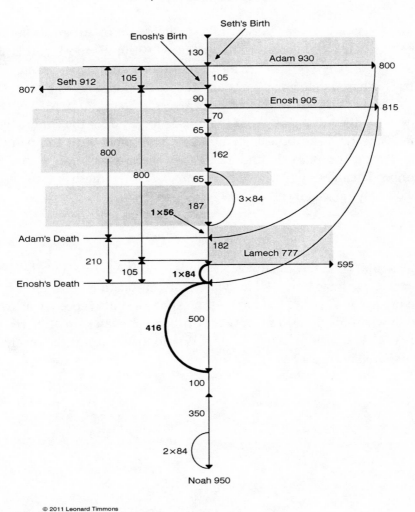

© 2011 Leonard Timmons

Figure 2.9: Enosh. The lifetime of Enosh is set based on an 800-year timespan bracketed by 105 years on each side. This method is very similar to the method used to set the lifetimes of Adam and Seth in Figure 2.7.

The interval between the death of Adam at the 56th year of Lamech's life and the death of Enosh at the 84th year of Noah's life is 210 years. The lifetime of Enosh is constructed by "sliding" the 800 years of Adam's life after the birth of his son Seth down to divide the 210-year period in half. Doing this creates a 105-year interval in Seth's lifetime before the birth of Enosh and sets the lifespan of Enosh to 105 + 800 = 905 years. This construction produces another 800-year interval bracketed by two time periods in a manner very similar to that used to set the lifetimes of Adam and Seth. See Figure 2.7.

Why use a 210-year period to link the death of Adam to the 84th year of Noah's life? Because 210 is a "nice," "round" number. It is also 21×10, and 21 is 7 + 7 + 7, which gives us another triple seven. This puzzle is based on triple sevens and the number of days in a year.

Note also that the puzzle authors reversed 56 to produce 65, which they used liberally throughout the puzzle to stand for the triple seven. This same type of numeric symbolism is possible with the pairs 84 and 48 and the pairs 21 and 12. It may seem a bit strange that the number 12 can be symbolic of the number 777, but these authors seem to be implying that it can.

2.3.2 Kenan

The authors fix the death of Kenan 416 years before the death of Lamech. We have to work backward from his death to determine how the puzzle authors set the moment of Kenan's birth. Since we already know when his father Enosh was born, once Kenan's death is set we can calculate that Kenan's lifetime could be a maximum of 1,000 years. See Figure 2.10.

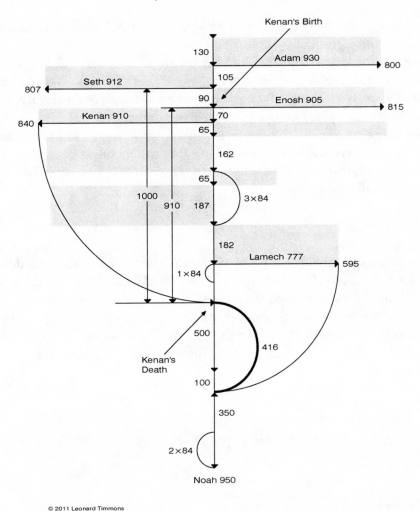

© 2011 Leonard Timmons

Figure 2.10: Kenan. The lifetime of Kenan is set arbitrarily but significantly within a 1,000-year span as 91×10 = 910 years.

It is very interesting that a 1,000-year span can be constructed at all with numbers that have to be related to the number seven and the Divisible Year. That this can be done is a testament to the skill of the puzzle authors.

So how did the puzzle builders set the lifetime of Kenan? In Section 2.2, we noted that (2×56) + (3×84) = 4×91 = 364, and that 77 + 7 + 7 = 91. The puzzle authors set Kenan's lifetime arbitrarily but significantly at 910 years. The number is arbitrary because many other numbers in the 900 range could easily have been chosen. The number

910 is significant because it is related to the 365 years of Enoch's lifetime through the Divisible Year.

Since the puzzle authors had set the time of Kenan's death, when they set his lifetime to 910 years, his father Enosh's lifetime was divided in two. Enosh is 90 years old when Kenan is born and lives an additional 815 years before he dies. The numbers 90 and 815 provide two more stumbling blocks over which you might stumble while trying to solve this puzzle. Also note that the lifetimes of Seth at 912 years and Enosh at 905 years are very close to 910, but are arrived at by *very* different means.

2.3.3 Mahalalel

The death of Mahalalel is set opposite to that of Kenan. Kenan's death is set 416 years *before the death of Lamech*, and Mahalalel's death is set 416 years *after the birth of Lamech*. Also associated with Mahalalel is the 3×84 = 252-year interval between the birth of Enoch and his grandson Lamech. To continue building the puzzle, the authors would have to provide a rationale for setting the lifetime of Mahalalel. See Figure 2.11.

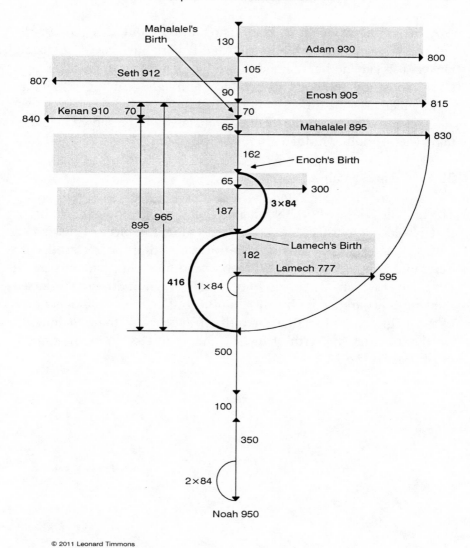

© 2011 Leonard Timmons

Figure 2.11: Mahalalel. The lifetime of Mahalalel is bounded by the 965-year span from the birth of his father Kenan to the 416th year of Lamech's lifetime. His lifetime is fixed at 895 years when his birth is set 70 years after the birth of his father, Kenan.

The puzzle authors set the lifetime of Mahalalel when they have his father, Kenan, live 70 years before his birth. Since Mahalalel's *death* was set to 416 years after the birth of Lamech and his *birth* has now been fixed, we know that his lifetime is 895 years long. So the authors set Mahalalel's lifetime by continuing the rationale they used to set Kenan's lifetime to 910 years. They simply removed 70 years from the 910 years of Kenan's life to get 840 years. The 840 is 84×10 and 84 = 77

+ 7, the number we have been using throughout this part of the puzzle. When Kenan's lifetime is divided in this way, the puzzle makers use yet another method to produce a number in the 900 range. Using so many methods to produce numbers that are so similar greatly increases the difficulty of this puzzle.

A second way to set the birth of Mahalalel will be shown below. It depends on another hidden reference point.

2.3.4 Jared and Noah

The lifetimes of Jared and Noah are related to one another. We know that Noah was born $182 = 364/2$ years after the birth of his father, Lamech. After setting this up, the puzzle authors were faced with determining how much longer Noah would live. They had already set the point at which Jared's son Enoch was born, and doing that caused the length of Jared's life before the birth of Enoch to be independent of the length of his life after the birth of Enoch. It is the remainder of Jared's life after the birth of his son that is related to the lifetime of Noah. See Figure 2.12.

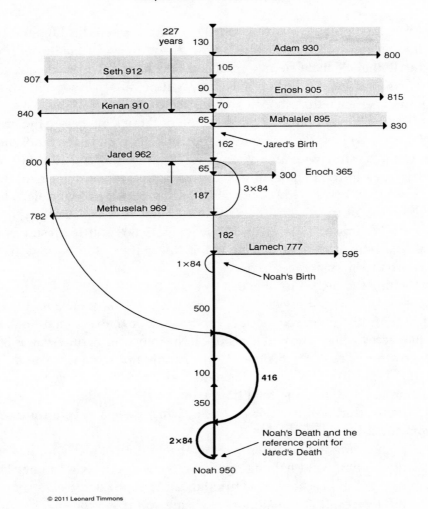

Figure 2.12: Jared and Noah. The lifetimes of these two patriarchs are related by a reference point at the end of the time line. When the length of Jared's life after the birth of Enoch is set to 800 years, Noah lives for 950 years.

To create the 1×84, 2×84, 3×84 sequence and to give the puzzle elegance, the puzzle authors inserted a 416-year span and a 2×84-year span into the time line to relate the two deaths. The lifetimes of Noah and Jared are then set to accommodate this requirement. The puzzle makers decided to set the death of Jared 800 years after the birth of his son Enoch, and choosing that value set Noah's lifetime to 950 years. So after the birth of Enoch, there is 800 + 416 + (2×84) years to the end of the time line. Measuring back from this time to the birth of Noah sets Noah's lifetime to 950 years.

The time of Jared's birth is unrelated to the rest of his lifetime. We know there is a 227 = (65 + 162)-year period between the birth of his grandfather, Mahalalel, and grandson Enoch, and the puzzle authors have to set Jared's birth somewhere within this period. We know from experience that choosing either the duration of Mahalalel's life before the birth of Jared or the duration of Jared's life before the birth of Enoch will be fairly arbitrary, but will fit within an established pattern, possibly with an inventive variation. In this case, the authors went back to the pattern of using a fixed time period with two time periods attached to either end. They set the period of Mahalalel's life before the birth of Jared to 65 years. Doing this sets the length of Jared's lifetime to 962 = 800 + 162 years. Both 800 and 162 result from fairly arbitrary choices that fit within the pattern the authors have chosen to follow in this puzzle.

Setting the period in Mahalalel's life before the birth of Jared to 65 years also causes a 65-year – 162-year – 65-year sequence to appear directly in the main sequence of the puzzle for the very first time. In other cases where this pattern is used, the intervening timespan is 800 years. In this case, the 800 years (the duration of Jared's life after the birth of Enoch) starts within this sequence rather than being contained within it. (Note that the total timespan of 65 + 162 + 65 = 292 days is 4/5 of the duration of a 365-day year, which seems to be a fortunate coincidence.)

Let me bring to your attention that the authors of these puzzles are clearly doubling and halving numbers to obscure their meaning. The number 162 is probably not important in this puzzle. It may simply exist as the result of the authors choosing so many other parts of the puzzle—that is, it is the "fix-up point." If we halve the number we get 81, which is 3 to the 4th power, and adding 3 and 4 equals 7. This is a fortunate outcome, but it does not seem to fit within the pattern of number play established by the puzzle authors.

2.4 The Third Part: Methuselah

The death of Methuselah completes this three-part puzzle. Methuselah's birth is set at the end of Section 2.2, where we discuss his father, Enoch. He is born when Enoch is 65 years old. We also know Methuselah's son Lamech is born when Methuselah is 182 years old. To fully define Methuselah's life, we need a rationale for setting the duration of his life after the birth of his son Lamech. In setting the lifetime of Methuselah, the puzzle authors returned to the method they used to set the lifetimes of Adam and Seth. See Figure 2.7. In that case, a 2×56-year period is placed at the end of an 800-year span, and a 2×65-year period is placed at the beginning of the span. The number 56 is hidden in the age of Seth as 800 + (56×2) = 912 years, and the number 56 is also hidden in the age of Adam by being reversed and then doubled to get 800 + (65×2) = 930 years. Every method used as a rationale for setting the lifetime of each of the other patriarchs is used in the production of Methuselah's lifetime. The puzzle authors use this device to provide the ultimate confirmation that we have found the solution to this puzzle.

When the lifetimes of Jared and Noah were set above, Hidden Reference Point 2 was created near the end of Noah's lifetime at the junction of the 2×84-year span and the 416-year span that set the death of Jared. See Figure 2.13.

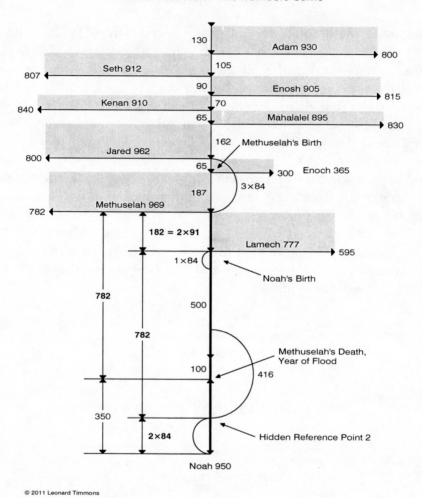

© 2011 Leonard Timmons

Figure 2.13: Methuselah. Only the duration of Methuselah's lifetime after the birth of his son Lamech has not been determined at this point. The determination of Methuselah's lifetime draws on all the other methods used to set lifetimes.

From Noah's birth to Hidden Reference Point 2 is 950 − (2×84) = 782 years. This 782-year span is flanked by a 2×84-year span at its end and a 2×91 = 182-year span at its beginning. The 782-year span is used to set the lifetime of Methuselah after the birth of Lamech. This construction is similar to the one used to set the lifetimes of Adam and Seth. See Figure 2.7.

The length of time from Hidden Reference Point 2 up to the birth of Noah is 782 years. To set the lifetime of Methuselah after Lamech's birth, the puzzle authors "slid" this period *up* to the birth of Lamech.

In the process they created a 350-year period at the end of the lifetime of Noah. By creating this construction, they made it clear that 350 = (2×84) + (2×91), and that this particular 350 years was a 168-year period and a 182-year period joined together. This method of specifying the duration of Methuselah's lifetime after the birth of Lamech is very similar to the way the lifetime of Enosh was set. In that case, they divided the 210-year span after the death of Adam into two 105-year spans by "sliding" an 800-year span *down* by 105 years. See Figure 2.9. In this case we "slide" the 782 years from Noah's birth to the hidden reference point *up* by 182 years, creating the 350-year span at the end of the time line.

The lifetimes of Kenan at 910 years, Enoch at 365 years, and Lamech at 777 years were chosen to be "nice" numbers. The lifetime of Methuselah was calculated so that it would also be a "nice" number, but one that we would not recognize as a "nice" number until we had the full solution to the puzzle. Once we have the full solution below, it will become clear why the authors thought 969 was a "nice" number.

Kenan's lifetime is 910 years long, and his lifetime after the birth of his son is 840 years. Each of these two numbers is used as a flanking time period in the lifetime of Methuselah. For Methuselah, the numbers 91 and 84 are the flanking numbers, and they are doubled to get 182 and 168, as shown in Figure 2.13.

Jared's lifetime before the birth of his son is set independently of his lifetime after the birth of his son. This is true of Methuselah as well. Because Jared's lifetime before his son's birth contains the "fix-up point" of the puzzle, this time must be set independently of his lifetime after his son's birth. In Methuselah's case, the authors have a specific reason for setting his lifetime before the birth of his son to 187 years.

Another important criterion for the lifetime of Methuselah is that he must outlive his son Lamech. This criterion was set to give the initial clue to the puzzle's solution. Lamech and Enoch are outlived by their fathers, and they are the focal points of this puzzle.

2.5 800 Years

The number 800 is very important in this puzzle. It is strongly associated with Adam, Seth, Enosh, and Jared. It turns out that 800 = 65 + (7×105). While this equation may not actually knock your socks off, it should. The puzzle generates this equation for us, and the means by which it is produced is shown in Figure 2.7 above, together with Figure 2.14 and Figure 2.15 below.

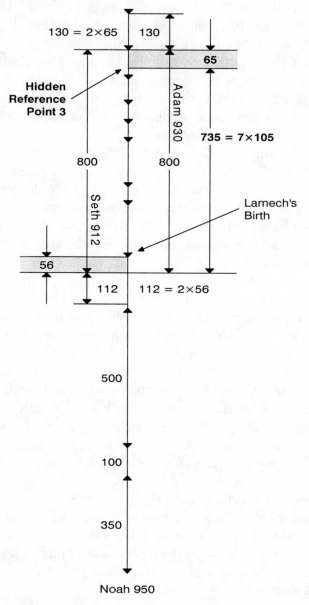

© 2011 Leonard Timmons

Figure 2.14: Symmetry. The shared 800-year period in the lifetimes of Adam and Seth lacks symmetry without a 65-year period at its beginning to match the 56-year period at its end.

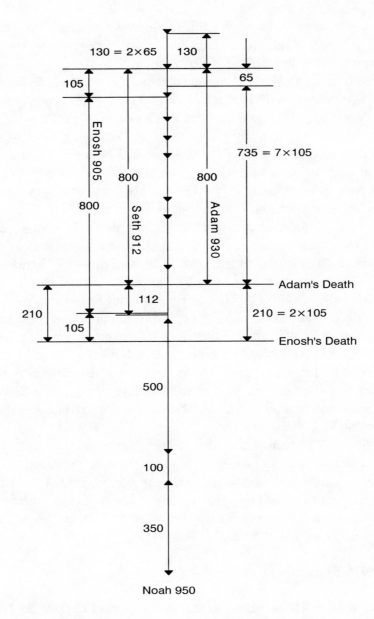

© 2011 Leonard Timmons

Figure 2.15: 210. The 210-year period that separates the deaths of Adam and Enosh follows naturally from the seven 105-year intervals that precede it. The 210-year period emphasizes that the seven 105-year periods are actually 3.5 210-year periods.

Figure 2.7 introduces us to an 800-year timespan bracketed by a 2×56-year span and a 2×65-year span. The reference point for these time periods is at Lamech's birth, which is 56 years before the end of the 800-year span. If a reference point existed 65 years after the beginning of the 800-year span, then this construction would be symmetrical. Marking the 65-year reference point at the beginning of the 800-year span divides it into 65 years and 735 years.

Now, here is where the arbitrariness comes in. When the puzzle authors were generating this puzzle, they wanted to choose a number so that a "nice," "round" number would result when added to 65. That number would also have to be divisible by 7. Only 35, 735, 1435, etc., would work—and only 735 would be large enough without being too large. So they chose to use the number 800 = 65 + (7×105). One of the consequences of constructing the number 800 in this way is that this construction makes a clear reference to 105 as half of 210. The number 210, as used in this puzzle, is symbolic of 777 and, as half of 210, 105 becomes symbolic of 777 as well. The net result is that 800 becomes symbolic of 777 in this round-about fashion.

I don't want to miss the opportunity to talk about the fact that 800 = 65 + (3.5×210). In this rendering, we have a multiplier that is equivalent to a "time, two times, and half a time." This 3.5 multiplier is used frequently in the Bible to represent a mysterious time period. That period could be 800 years or 800 days, if the 3.5 in this equation is used to symbolize the number 800. It could also symbolize the number 777.

Figure 2.15 shows how the puzzle authors connected the first and second parts of the puzzle through the use of the numbers 210 and 105. The authors "slide" the 800-year period shared by Adam and Seth down to divide the 210-year period in half. The authors use this construction to connect the first and second parts of the puzzle and to give us a clue that the number 800 is not completely arbitrary.

2.6 416 Years

The number 416 is central to the second part of the puzzle. Numerical symbology is also associated with its use. This puzzle is about the numbers 7 (mostly as 777) and 364—the number of days in a Divisible Year. We know that 500 = 416 + 84, which is a really great result since 500 is a "nice," "round" number, and 84 is symbolic of 777 since 84 = 77 + 7.

The number 416 can be symbolic of 777 and *also* symbolic of the Divisible Year. This is done in the following fashion: 56 = (7×7) + 7 and $416 = (7\times52) + 52$. The number 56 is directly symbolic of 777, and $416 = (7\times52) + 52 = 364 + 52$ is calculated in the same fashion, but uses the Divisible Year. So the number 416 is symbolic of a year without being a year in length.

Now, a 364-day year is approximately 1.25 days shorter than a true solar year. So every four years, five days must be added to the calendar to keep the calendar in sync with the seasons. If you look at Figure 2.8, you will note that exactly four 416-year periods are in this puzzle. These periods represent four years of our synchronized calendar. The five days that must be added every four years are shown as the five-year gap between the death of Lamech and the death of Methuselah, who dies in the year of the Flood (see Figure 2.4 for more detail).

The clear implication here is that there could have been a secret calendar based on this mechanism. If this were true, then that calendar should appear somewhere in this puzzle. And it does. See Figure 2.16.

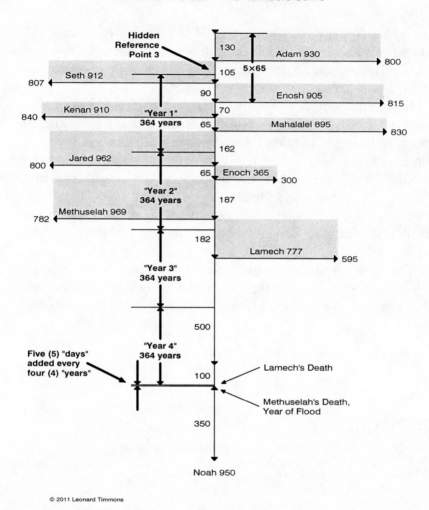

© 2011 Leonard Timmons

Figure 2.16: Divisible Calendar. This puzzle contains a calendar based on the Divisible Year. Each year is 364 days long, with an extra five days added every four years to keep the calendar in sync with the seasons. The calendar begins at Hidden Reference Point 3 of Figure 2.14, which occurs 65 years into the lifetime of Seth.

The Divisible Calendar begins from Hidden Reference Point 3 in Figure 2.14. That hidden reference point results from the need for symmetry in the relationship between the lifetimes of Adam and Seth. From that point to the death of Lamech is 4×364 years, and that time period represents four 364-day years. After that, five days are added to the four years to keep the calendar in sync with the seasons. This is an amazing result and is the focal point of this puzzle.

This new calendar is synchronized with the seasons, but it can be improved. Better synchronization can be obtained by adding a total of

41 days to a 364-day-per-year calendar every 33 years. Does that calendar exist in this story? It sure does. If we use eight four-year cycles, each of which contains five intercalary days, then we would add an extra 40 days in 32 years. If we make the 33rd year exactly 365 days long (add one intercalary day), then the 33-year cycle would very precisely maintain the relationship between the calendar and the seasons. In this puzzle, the 33-year calendar is constructed using the story of Noah and his ark.

Noah built an ark to preserve his family and the air-breathing, land-dwelling creatures of the earth from a massive flood. The Flood lasted a little more than a year. Eight people including Noah entered the ark on the day of the Flood. The eight people represent the number of times to repeat our four-year calendar. Then, to complete the calendar, an extra year—the year of the Flood—is added to the calendar. This is the 600th year of Noah's life. In addition, this 33rd year is 365 days in length, as indicated by the passage "Noah walked with God," which is the same statement used to describe Enoch, who lived exactly 365 years. The length of this final year is not indicated by the number of days and months of the Flood story, since the Flood does not last for exactly 365 days.

I think it is clear that this puzzle encodes a highly accurate solar calendar designed to maintain the relationship between the calendar and the seasons. The encoded calendar uses a Divisible Year of 364 days as the common year. A Leap Year occurs every fourth year and contains five intercalary days, making it 369 days long. The entire four-year period is 1,461 days long. Eight of these periods are then concatenated, and a single 365-day Perfect Year is appended for a total of 12,053 days in 33 years. The average length of the year produced by this procedure is 365.242424 days.

For a full discussion of the many different ways to measure the length of a year, please see Appendix B in the book *Marking Time: The Epic Quest to Invent the Perfect Calendar* by Duncan Steel. He reports the mean length of the year from spring equinox to spring equinox[7] to be 365.242374 days.[ah] This year length is very close to the length we have calculated above and is the year length that remains the most

---

7   We have two equinoxes, one in the spring and another in the fall. At the equinox, the number of daylight hours is equal (*equinox*) to the number of nighttime hours. The spring equinox is called the *vernal equinox*, and the year measured from spring equinox to spring equinox is the *vernal equinox year*. Our calendar is based on the average length of this year.

stable over time.[ai] It also bears mentioning here that most calendars are *not* based on the mean tropical year of 365.2422 days, and this misunderstanding is widespread.[aj] I was unaware of calendars based on the 33-year cycle before I discovered the relationship above. I have since found two: the Jalali calendar and the Persian calendar.[ak] It is clear from our discovery here that the 33-year calendar was known at least many hundreds of years BCE[8] depending on when you believe the book of Genesis was written down or later edited.

As an aside, some interesting arithmetic gymnastics are possible here that don't really fit the pattern established by the authors. With 41 days added to the calendar every 33 years, a convenient way to memorize the formula is to concatenate the digits to get 4133. If you add the two lower digits you get 416, which is probably just a coincidence. In this same vein, however, note that 416 is 56 if you add the digits 4 and 1. It is tempting to say that 416 was chosen for all these reasons because it would be hard to imagine that these authors were unaware of these associations. Even though the puzzle makers are clearly playing with these numbers (it's a numbers game), I don't see any evidence that the authors chose these numbers for these reasons.

Hidden Reference Point 3 from which the calendar begins in Figure 2.16 (and Figure 2.14) is placed at an interesting point. It divides the 105 years of the lifetime of Seth before the birth of his son Enosh into 65 years and 40 years. The 40-year period is followed by the 90 years of Enosh's lifetime before the birth of his son. Together, the 40-year period and the 90-year period create another 130-year period. So from the birth of Adam to the birth of Kenan is 5×65 years. See Figure 2.16 above. The entire period from the birth of Adam to the birth of Kenan also displays the repeating pattern of a specific time period flanked by two time periods of equal (or related) length. In this case, the central time period is 65 years flanked by two 130-year periods. When we consider the other two 65-year periods in the mainline of the puzzle, there are seven 65-year periods in all.

Another numerical relationship is also apparent in Figure 2.16. With a little calculation, you will note that the middle of this four-year cycle occurs 49 years after the birth of Lamech. Since Lamech lived 777 years, we have 777 = 49 + 364 + 364. This is a very nice

---

8   Before the Common Era (BCE). The Common Era (CE) is equivalent to AD, and BCE is equivalent to BC.

relationship from which we can set the date of Lamech's birth. Lamech's birth date is the most important date in this puzzle.

2.7 369 and 187 Years

The years of the Divisible Calendar come in three lengths. A Divisible Year is 364 days long. A Perfect Year is 365 days long. And a Leap Year is 369 days long. The 364-day Divisible Year is hidden in this puzzle in spectacular fashion, as shown in Figure 2.16. The 365-day Perfect Year is not hidden at all—it is the number of years in Enoch's life. Since these numbers are important parts of the puzzle, one might also expect the puzzle authors to include the number 369 as well. And they do. See Figure 2.17.

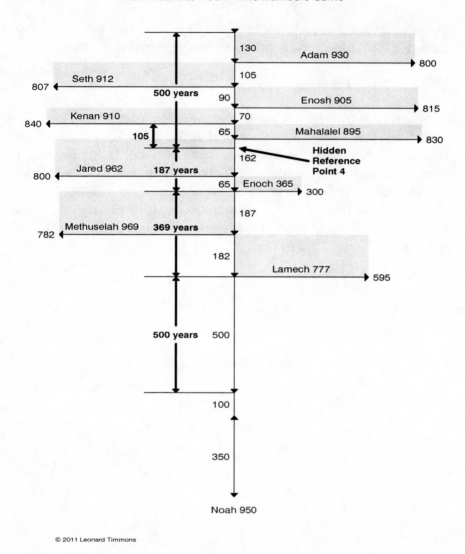

© 2011 Leonard Timmons

Figure 2.17: Overall Structure. The overall structure of the puzzle is shown here. Fundamental to this structure are the Leap Year and the Half Leap Year that form a single period 369 + 187 = 556 years long. The pattern of flanking this period with two periods of equal length uses 500 years in this case.

From the birth of Methuselah to the birth of Noah is 369 years. This time period is divided into a 182-year period, which is clearly one half year and a 187-year period that is half a year plus 5 days. Now we know the secret of 187 = 182 + 5. The Leap Year is 369 = 182 + 187 days long and is divided into a half Divisible Year of 182 days and a Half Leap Year of 187 days. This division of the Leap Year into 182 and 187 days is the reason why 187 years is selected as the period in Methuselah's lifetime before the birth of Lamech. So ultimately, Methuselah's lifetime of 969 years is 369 + 600, which makes it a combination of a "nice" number and a "round" number.

Figure 2.17 shows the overall structure of the puzzle. It is interesting that the Leap Year plays such an important role, but one must first know that a Leap Year is 369 days long before the number makes any sense. One must also know that a Leap Year is not divided in half, but is divided into 182 days and 187 days. The authors also carry through their pattern of appending two equal time periods to the ends of a central time period. In this case, the central time period is 369 + 187 = 556 years long, and each end is 500 years long. In terms of "sliders," a time period of 1,056 years would be "slid" by 500 years to produce the three main time periods. The center time period of 556 years would then be subdivided into 369 and 187 years.

As you can see from this figure, Hidden Reference Point 4 is created within the lifetime of Jared before the birth of his son Enoch. The reference point marks a time 40 years after Jared's birth. The authors set the birth of Mahalalel at 105 years before this reference point. This is the second way that the first and second parts of the puzzle are linked using the number 105. Setting the birth of Mahalalel in this way explains why he is born 70 years after his father Kenan was born.

So the 500 years at the beginning of the time line is divided into a 105-year period preceded by a 70-year period, which is then preceded by a 5×65 = 325-year period. In addition, the 187 + 105 years between the birth of Mahalalel and Methuselah is divided into 65 + 162 + 65 years. And with this overview, it becomes absolutely clear that the 162-year period really is the "fix-up point" of the puzzle.

Another interesting thing about the Leap Year length of 369 days is that it is also divisible. More interestingly, one of its divisors is 41—the number of days added to a single 33-year cycle. So we have 369 = 41×9 or 369 = 123×3. Since this is a numbers game, these authors would think that having three 123-day periods in a Leap Year would be an extremely nice result. In addition, a four-year cycle contains 1,461 days. Our puzzle authors surely would have found the very interesting relationships that result from playing with these two numbers. For instance, 1461 − (10×123) = 231 is interesting, since it rearranges the 123 to get 231. Also available in this series is 1461 − (7×123) = 600, and we get the 600 "years" that when added to 182 years give us the 782 years of Figure 2.13 for the duration of Methuselah's life after the birth of his son Lamech. Note also that Methuselah's entire lifespan is described by the equation 1461 − (4×123) = 969 and 600 + 369 = 969, as shown in Figure 2.17. (Since 1461 contains a 3×123 component, the following are also true: 3×364 = 1092 and 1092 − (7×123) = 231, 1092 − (4×123) = 600, 1092 − (1×123) = 969.)

What we are doing with this sequence is taking one 123-day period from three 364-day years (969), one 123-day period and a Leap Year period (369 days) from three 364-day years (600), and one 123-day period and two Leap Year periods (369 days) from three 364-day years (123). That's the pattern. And as a final point of interest, 1461 − (9×123) = 354, which is the approximate length of a 12-month lunar year—a result that must have seemed somewhat mystical to the authors of this puzzle.

Since the number of leap days added to a 33-year cycle (41) is a sub-multiple of the number of days in a Leap Year, it is easy to calculate how many 33-year cycles it would take to have added a full Leap Year of extra days to the calendar. Nine 33-year cycles would be required to add 369 leap days to the calendar, and 9×33 is 297 years.

Another piece of information that's fairly easy to calculate is the number of years that would have to pass for our seven-day weekly cycle within the calendar to repeat itself exactly. The answer is simply 33×7 = 231 years. We get this same number above: 231 = 1461 − (10×123) = 354 − 123. (This equation also makes it clear that the lunar year of 354 days is the sum of 123 and 231, a result that also might seem mystical.)

2.8 What the Puzzle Enshrines

We have no evidence that this puzzle has been understood by anyone for at least 2,500 years, and maybe much longer. Decoding it is an archaeological discovery of some importance. I think it compares to other major archaeological discoveries. It's like discovering the unopened tomb of King Tut and being the first to look inside. Exploration of the tomb gave us information that allowed us to look further into the hearts and minds of the people who built it. Similarly, now that we've discovered a calendar in Genesis 5, we know much, much more about the people who wrote it. Opening Tut's tomb was only the beginning of the discovery. Now, we've just discovered a fully functional calendar in Genesis 5, and in the entirety of Genesis 1 to 11, many more discoveries are waiting to be made.

This puzzle is an absolutely wonderful thing. Its rules, structure, and purpose are a delight to think about. I made the first discovery just after Christmas, 1999. It's been a true pleasure to work with since then. I've discovered more and more about it, and I'll show you those additional discoveries below. Over time, I've come to believe that I've figured out most of it, though I certainly could have missed something important.

This puzzle is a work of art. The skill required to create it is clearly extraordinary, even by today's standards. The Bible is filled with literature, much of which was created as works of art. Examples of this are the Song of Solomon and the Psalms of David. This work is so precise that it brings to mind the drawings that might be used to build a beautiful structure like the Golden Gate Bridge. This is high art and high technology.

This puzzle is also a work of science. It encodes one of the most important scientific documents of the ancient world, the calendar. More importantly, this calendar is as precise as any we have today. These ancient scientists had to determine the length of the year precisely. Complicating that determination is the fact that each year *varies unpredictably* in length from the year before. The *average* year length measured from spring equinox to spring equinox[9] is stable, however. So these ancient scientists had to determine the average year length. Collecting the data and making the calculations required years of determination, organization, and the means to keep time accurately. The people who measured the average year length had significant technology and extraordinary intelligence. We must not underestimate the technology and intelligence needed to precisely determine the length of the year. These people were more than very much like us—they *were* us. While human culture (what we do) and human knowledge (what we know) may have changed over the last several thousand years, humankind (what we are) has not.

One of the more interesting things this puzzle does is to provide the basis for most, if not all, of the numerology in the Bible. The Bible authors scattered numbers throughout the Bible, and in many instances those numbers seem to have special meaning. If you've ever spent time trying to understand them, it's mostly time wasted. The origin of the numbers is unknown, and the writers seem to expend

---

9   The vernal equinox year.

some considerable effort to use them in ways that obscure their meaning. So, for instance, when we read the story of Jesus' appearance to his disciples after his death, we get the following:

> After this Jesus revealed himself again to the disciples by the Sea of Tibe'ri-as; and he revealed himself in this way.
>
> Simon Peter, Thomas called the Twin, Nathan'a-el of Cana in Galilee, the sons of Zeb'edee, and two others of his disciples were together. Simon Peter said to them, "I am going fishing." They said to him, "We will go with you." They went out and got into the boat; but that night they caught nothing.
>
> Just as day was breaking, Jesus stood on the beach; yet the disciples did not know that it was Jesus. Jesus said to them, "Children, have you any fish?" They answered him, "No." He said to them, "Cast the net on the right side of the boat, and you will find some." So they cast it, and now they were not able to haul it in, for the quantity of fish. That disciple whom Jesus loved said to Peter, "It is the Lord!" When Simon Peter heard that it was the Lord, he put on his clothes, for he was stripped for work, and sprang into the sea. But the other disciples came in the boat, dragging the net full of fish, for they were not far from the land, but about a hundred yards off.
>
> When they got out on land, they saw a charcoal fire there, with fish lying on it, and bread. Jesus said to them, "Bring some of the fish that you have just caught." So Simon Peter went aboard and hauled the net ashore, full of large fish, a hundred and fifty-three of them; and although there were so many, the net was not torn. Jesus said to them, "Come and have breakfast." Now none of the disciples dared ask him, "Who are you?" They knew it was the Lord. Jesus came and took the bread and gave it to them, and so with the fish.
>
> This was now the third time that Jesus was revealed to the disciples after he was raised from the dead.[al]

In this story, it is odd that the number of fish caught was important enough to mention. Why would anyone care, in the whole scheme of things, how many fish were caught? But if you look closely, you will notice that there are 153 large fish and that $153 = 77 + 77 - 1$. Even more interestingly, if you note that Jesus is cooking a large fish, then there are $154 = 77 + 77$ fish in total.

So what is the meaning of double and quadruple sevens? What is the significance of the triple sevens in this puzzle? These odd numerical relationships continually pop up throughout the entire Bible. Since our puzzle is the first occurrence of these relationships, and since they are presented in such a spectacular fashion, I believe that much of the numerology of the Bible traces back to this puzzle and to this calendar.

2.8.1 + 0 Years

Take a look at Figure 2.8 again. Notice the four time periods that are 416 years long. Three of them are associated with periods that are multiples of 84 years. We can represent the four of them as follows:

$$416 + (0 \times 84)$$
$$416 + (1 \times 84)$$
$$416 + (2 \times 84)$$
$$416 + (3 \times 84)$$

We know that multiplying a number by zero produces a zero result. If these authors knew that, it means they understood the concept of zero much earlier in history than originally thought. I think they did. In the first part of the puzzle, the authors created the sequence 1×56, 2×56, and 3×56 and associated each element of the sequence with the death of a patriarch. No association of 0×56 was made with the death of a patriarch, so we cannot reasonably infer that the authors would represent the starting point as 0×56. However, in the case of a series of disconnected line segments that form a sequence, we can reasonably infer that the authors understood that multiplying a value by zero produces a zero result.

We don't really know how old this puzzle is, but depending on its age, these authors may have understood the concept of zero. If so, it had to be such an important and valuable concept that it deserved a place within this most important of puzzles.

2.9 Hiding in Plain Sight

A calendar hidden in Genesis 5 is difficult to believe without proof. People have been looking at these numbers for thousands of years, and no one has noticed that the numbers were hiding a calendar. Since I'm claiming that a calendar can be found in a place where almost no one expects to find one, I am obligated to prove that the calendar is there. I think I've done that here.

One of the things we must accept at this point is that the Bible writers *did not want us to find the calendar*. They hid it so well that it was almost impossible to find. Why? And why did I find the calendar when so many others failed to, even though it was hiding in plain sight? My answer is that I was willing to put in the time to figure out what the numbers meant, and I was willing to look at the world through the eyes of the people who wrote the Bible. I was trying to look at the Bible through the eyes of its writers long before I solved this puzzle. It was this willingness that allowed me to solve the riddles of Noah's ark, the Garden of Eden, and Revelation 13. I found the solution to those riddles first, and that allowed me to almost feel the solution to this one.

I was not the only one to imagine that something was hidden in these numbers. The problem with investigating the meaning of this genealogy is that so many people from the fringe are also investigating it. You put yourself in their company when you pursue this elusive goal. It's like being a UFO researcher. If you were a scholar and suspected that something was hidden in these numbers, pursuing that hunch would be a very good way to destroy your credibility and career. Yet some have tried to solve the puzzle and have characterized what a real solution must look like.[am] The solution I offer here is the one everyone has been seeking. But this is not the end of the matter. This puzzle does not stand alone. It is intertwined with the story of the Flood, and solving it forces us to look again at the entirety of Genesis 1 to 11.

The solution you see in front of you is the result of a process. As I became willing to put aside what I was told the Bible said, I started to solve some riddles and puzzles. With each new solution, I was forced to think differently about the book. This was not easy and it took a very long time to do. Eventually, I discarded what I *wanted* to believe about the Bible. I stood back and looked at the book without emotion, coldly, dispassionately, but with overwhelming force. I stopped myself from reading the Bible with the intent of confirming my beliefs about its writers or to have its writers agree with the beliefs I held. I began to read the Bible with the intent of having the authors tell me who they were. This puzzle, more than anything else, forced me to conclude that the authors of this puzzle were just that—puzzle makers.

During the more than one thousand years that it took to write the Bible, a persistent "wisdom culture" existed in the ancient near east that valued the creation and sharing of puzzles and riddles with hidden meaning.[an] We use the word "parables" to refer to these puzzles and riddles, but I will avoid doing that here because at this level of complexity, these riddles do not fit our modern definition of the word. The story that the Dodonaean priestesses told about the origin of their place of worship is an example of how this wisdom culture created one of its riddles. That story was built by the Dodonaean cult to be understood only by those "in the know."

This wisdom culture filled the Bible with "wisdom literature." The Proverbs of Solomon are a prime example of that literature. In the very first chapter, Solomon tells us directly and without equivocation that the wise speak in riddles. Anyone who considered himself to be a wise person would consider a beautiful riddle that held valuable knowledge as something to be desired—something to be created. Many of the works in the Bible were added to that collection because they were written by the wisest men of their time. For these reasons, the Bible has to be full of riddles. So the phrase "wisdom literature" must describe the first 11 chapters of Genesis[ao] and much of the remainder of the Bible.

Some of the proverbs are numerical in nature. The existence of these numerical proverbs tells us that playing with numbers was a part of this wisdom culture. So when we find that these authors were so enthralled by the relationships between the numbers in their calendar that they encoded that calendar and those relationships within this puzzle, we know that their actions were entirely in keeping with their culture.[ap]

As we step back and look at this puzzle and its solution, it's fairly easy to notice that the authors packed as many relationships into this puzzle as they could. It's clear that what they were doing was not accidental. This layering of complexity upon complexity was not done just to prevent us from discovering the calendar—it has a meaning of its own. In fact, this may be the most important piece of information in this puzzle. But we have to know what the Bible is before we can understand why the puzzle makers built their puzzle in this fashion.

CHAPTER 3

What is the Bible?

3.1 Insight Into the Text

THE BIBLE WAS WRITTEN over a very long time by many different people who shared a wisdom culture that remained relatively stable over that period. That culture has been lost, and our modern culture doesn't have anything that corresponds to it. We don't normally hide critically important information in complex stories that are incredibly difficult to understand. Most people would consider that kind of behavior to be annoying or ridiculous. The calendar story insists that these writers thought hiding critically important information in plain sight had both utility and beauty. So the discovery of a calendar in this story tells us that we have not understood this section of Genesis and, by implication, we haven't understood much of the Bible. I think that the depth of our misunderstanding is extreme. We don't know the ultimate intent of the writers. We don't know how many Bible stories are actually puzzles or riddles, like this story. We don't know what these writers thought was valuable enough to hide in a riddle, and we don't know what was common enough to remain unhidden. Our culture doesn't use riddles and puzzles as its fundamental way of dispensing real knowledge, so it is incredibly difficult for us to understand the Bible writers, their writings, and the people living in their world.

In our world, we are settled, and we sustain ourselves by growing our food, both plant and animal. Thirty-five hundred years ago, nomads were still plentiful. Hunting and gathering had been an important way of life, and its influence on cultures in western Asia was still intense. Gardening was a way of life, and farming was still advancing.[10] A worldview anchored in this environment produced a system of education that was totally unlike the one we know. Today,

---

10 This is gardening in the sense of caring for the area in which you live. A gardener would enhance the growth of desirable plants and prevent the growth of those that are undesirable. A farmer would purposely control the entire life of a plant including where, whether, and how it grows.

we teach our children from first principles. We teach them bit by bit throughout their school years. When the best of them have reached the highest level of education, we require that they do some original work that shows *insight* into their chosen field. These are the PhD candidates and their dissertations.

A system of education like ours did not exist 1500 years BCE. And even though many people were settled, the thought processes of the authors of the Bible were still influenced by their nomadic and hunter-gatherer origins. Just as in our system of education, they wanted to produce insightful people. We end with insight, but they *began* with insight. They thought of *insight as a characteristic* of a person. You couldn't give it to someone who did not have it. When viewed in this way, insight would be similar to an undiscovered talent. You either have it or you don't.

A person could have a talent for singing; he could also have a "talent" for insight. If you had a talent for singing, singing lessons would probably enhance your inborn talent. If you did not have a talent for singing, you could still take lessons if a teacher was available. The very best teachers might refuse to take you on as a student if, in their opinion, you did not have that inborn talent. That teacher might believe that without the inborn talent, you could never reach a high level of proficiency, so giving lessons to you would be a waste of their teaching talent. The Bible writers thought of their potential students in this same way. Their view was that a person could acquire insight if he had the capacity, but insight couldn't be imposed on him or given to him. These views of insight fit very well into a hunter-gatherer mentality, where one would search for food or any resource that grew of its own accord. These views of education also fit well into a gardener mentality where one would tend to a plant that grew of its own accord so that it might be even more productive, but one would not plant the seed. So an ancient "hunter-gatherer/gardener" educator would follow his instinct to search for a student who had the innate capacity for insight before attempting to teach him.

We might think a system of education that sought to find insight and enhance it would be inferior to ours. That conclusion would be incorrect, in my view. When applied in this manner to education, a hunter-gatherer/gardener view of the world should be quite efficient at directing very limited resources (teachers) toward those who might benefit most. I think such a system would be more likely to locate and encourage child prodigies and empower them to use their great insight to everyone's advantage.

In our system, the extraordinarily insightful people are the research scientists who create new knowledge and allow us to expand our understanding of the world. In this ancient system the extraordinarily insightful people were a part of the religious system. Prophets like Moses and Elijah created new knowledge and allowed the people of their time to expand their understanding of the world through direct revelation from God. The ultimate goal of our system of education and this ancient system are the same: to know and understand everything about the world. We create research scientists who do this work. To do the same work, this ancient culture sought out those with prodigious insight and allowed them to become the prophets of their time.

As a system like this evolved, it would have had to answer the question, "How does one test for insight?" The answer to this question determined the form of this ancient system of education and influenced everything produced by it—including, and especially, the Bible.

Insight is defined as:

1. The capacity to discern the true nature of a situation; penetration

2. The act or outcome of grasping the inward or hidden nature of things [...]ᵃ۹

To determine whether someone has insight, you would have to see them discern the true nature of a situation (or the inward nature of a thing) when that true nature is hidden. A great way to test that ability is to produce a riddle and to hide the solution. Should someone come up with the solution, then that person has insight into the true nature of the riddle and, more importantly, that person possesses insight itself (the talent, that is).

The amount of insight a person has can be measured by the difficulty of the riddle he or she can solve. Simple riddles are for those who have less insight and are appropriate for the simple-minded and children. Hard riddles are for those who have prophetic insight. Hard riddles can imitate the complexities of the world at large, since the way almost everything in the world actually works is hidden and extremely hard to figure out. For example, it is sometimes easy to know *what* is happening in the world, but it is often extremely hard to know *why* things happen the way they do. It's comparatively easy to find out that there are 365 days in a year, but to find out that there are exactly 365.2424 days from spring equinox to spring equinox (on average) is hard. The extremely hard thing to know is why there are exactly this number of days on average and why the length of a year varies slightly from year to year. To know that, you need a fairly good understanding of celestial mechanics, and that is beyond most of us, even today.

If you were an ancient hunter-gatherer/gardener educator trying to find students for your school, you would have to widely publish both simple and complex riddles and determine who in the population had the innate ability to solve them. The simplest riddles would be what we've come to know as parables, and the hard riddles would be those like the genealogy calendar above. The individuals who could solve the simple riddles would be candidates who could be taught how to enhance their talent for insight. But they might not be the best students. In order to find the best students, our hunter-gatherer/gardener educator would have to publish riddles that didn't seem like riddles and puzzles that didn't seem like puzzles. The best students would be those who first recognized that there was a puzzle to solve. Over time these stories would become a collection of stories that would be told, retold, and ultimately written down. The hunter-gatherer/gardener educator would establish a school that would, over time, become a school like one of Elijah's schools of the prophets. Jesus' "fishers of men" movement is another example of this kind of search for students.

Educating a person to be more insightful is not the kind of education needed for everyday living. The purpose of these "insight schools" would be to mentor those who could significantly increase our knowledge and understanding of the world. The purpose of what we could call the "standard schools" would be to pass knowledge from one generation to the next and help people live each day.

Both types of schools are needed, but conflicts often exist between the two. Everyone needs to live day to day. But the need to discover new knowledge can be viewed as a luxury, not an absolute necessity. The "standard school" would have this view. But deep insight is needed to successfully handle the unusual threat, to handle big jobs like ruling a nation, or to forecast an event that could result in the total destruction of a people. The insight school would tend to view these events as more important than survival from day to day, since that activity is futile if ultimate survival cannot be guaranteed. This conflict runs deep in societies. The schools are often in mortal conflict, and one or the other rules the day. When it was in control, the "insight school" was able to produce much of what we read in the Bible.

This is a real conflict. Should all of a nation's resources be consumed in the hope of producing a few people of great insight, or should those resources be dedicated to the real needs of real people whose needs are immediate and apparent? For the insight school, the ultimate future depends on insight, so the need to prepare the way for the "greatly insightful persons," the great prophets, is overwhelming. They were willing to put this entire system in place to detect those people when they appeared. The hard riddles not only allowed the teachers at the time to detect those with the insight talent, but those riddles helped enhance their students' innate ability to be insightful.

Hard riddles were an excellent means to preserve what was known and communicate it only to those who had insight. Clearly there was the expectation that those with insight would use that knowledge wisely. The insight school did not care about those who had no insight. Their purpose was to produce individuals of great insight. So the need to preserve the riddles from generation to generation would be very important to them. Likewise, a vocal or scribal tradition based on transmitting the riddles unchanged from generation to generation would be very important in accomplishing that task. To accomplish all of this, the insight school created and preserved a book containing a collection of hard riddles that has allowed its authors to communicate the important information of their civilization across thousands of years. That book would be both a library and a textbook. Not a textbook like any we've ever seen, but a textbook nevertheless. The Bible is that book.

The Bible is the primary textbook of this ancient system of education. As a textbook, the Bible's primary purpose would be to pass on the knowledge of the current generation to the next. We never recognized the Bible as a textbook because it teaches with riddles and puzzles. Since this method of teaching is so foreign to us, we could not understand what the Bible was, and up 'til now we haven't understood what it was trying to do. When a Bible reader has an insight that allows him to understand a riddle, or an insight that allows him to put a puzzle together, only when he has the solution in hand will he know what the previous generation wanted to communicate to him. In this way, the Bible communicates the critically important knowledge of its culture. When we look at the Bible like this, we understand that the Bible is an *incredibly important* and *extremely valuable* compilation of the knowledge of this civilization. These teachers built the Bible in this way to protect the most valuable of their assets from the ravages of time. From beginning to end, the Bible is composed of riddles, many of them hard riddles. One of those hard riddles has been solved above. We know the solution to that riddle, we know the valuable information contained within it, and we have extracted that information several thousand years after the riddle was built.

A perfect calendar was an extremely valuable thing to these authors, and we know that they stored that calendar within their "textbook Bible." I've found that the Bible contains objects of even greater value. If you read and think carefully about Genesis 1 through 11, it soon becomes clear that the writers were dealing with the fundamental questions that humans have busied themselves with since the dawn of history. Within these stories, the writers have encoded a philosophy. This is a view of the world that they wanted us to discover, but for the reasons mentioned above, they would not talk about it plainly and openly without using a riddle or a puzzle. Their philosophy covers subjects like the way the world works, how it came to be, what it is composed of, how one thing affects another, what the purpose of our lives should be, and what we should ultimately do with our lives. I think I've discovered what that philosophy is through my study of the Bible over my lifetime.

As I mentioned above, we must always be careful not to force our view of the world onto those who wrote the Bible. To truly understand the book, each of us has to try to see the world through the eyes of the people who wrote the Bible at the time they were writing or editing it. I don't think I could have discovered the solution to these puzzles and riddles without allowing myself to think like the Bible writers thought. To make myself think like they did, I had to face the fact that people are the same now as they were then. We're not more advanced than they were. They were not more primitive than we are. People have not changed. What has changed is our culture and the way we look at the world around us. Their worldview is embodied in the philosophy that they built and skillfully hid within the text of the Bible. That philosophy allowed them to accurately model and explain the way their world worked.

When we talk about things like minds, souls, angels, spirit, or gods, we must know that these concepts are very nebulous. For so many, an angel is a man-like creature with wings. We think that souls and spirits are somehow related, but often we're not quite sure how. We're sure that we have a soul, and we can feel it, but we can't quite define it. We're also certain that we have a spirit and that it is more ethereal than a soul—or maybe it's the same thing. We know that a spirit can be broken, but are not sure how that spirit is related to a spirit of compromise. The Bible writers have clear definitions for the terms they use, and those definitions can be extracted from their writings. You will very likely disagree with these definitions. They may be difficult to accept because you have your own definitions that you've been comfortable with your whole life. The definitions you will find here are consistent with the Bible and with one another. They allow a real philosophical discussion to take place.

Like philosophers today, the Bible writers were focused on how knowledge works. Their understanding of the way knowledge works starts with insight, continues with knowledge and understanding, and concludes with wisdom. Understanding the interplay among insight, knowledge, understanding, and wisdom is important and allows us to understand the philosophy of the Bible writers. We have grossly misunderstood their philosophy because we did not know what we were reading when we picked up the Bible and began to read.

3.2 Knowledge and Understanding

Knowledge is at the foundation of this philosophy. Knowledge is simply an accumulation of data. We accumulate data from our senses. Knowledge results from directly sensing the world around us. Understanding is a particular kind of knowledge. Understanding is *knowing the relationships* within the data you have accumulated (that is, the knowledge that you have).

If you understand something well enough, you can make accurate predictions about it. Let's say that you understand cars and drivers and crashes. If you saw two cars driving directly toward one another and both drivers refused to turn aside, you could predict that the cars would crash. If you knew the speed and location of each car, you could also predict where and when they would crash. You know precisely what is going to happen before it happens. Foreknowledge is knowing what is going to happen before it happens. As your understanding of the world increases, your ability to know in advance what will happen in the world also increases. High-quality understanding can allow you to have foreknowledge of a situation that is *just as good as directly observing that situation*. You could have foreknowledge of a thing that is *just as good as directly sensing that thing*.

Knowledge and understanding seem to be different, but they are, in fact, the same kind of thing. Understanding is a type of knowledge. Understanding is *knowing* the relationships that exist within the raw data. Relationship knowledge (understanding) is not directly sensed from the world, but is produced by the process we've been calling insight.

Relationship knowledge is what is known as meta-knowledge: that is, it is knowledge that characterizes other knowledge. Because a relationship is also data, it is possible to find relationships within collections of relationship data. This *meta*-meta-knowledge is a "higher" level abstraction of the data and represents a "deeper" understanding of the original data we received from our senses. This process of abstraction can continue indefinitely unless an Ultimate Meta-knowledge Relationship (UMkR) or Ultimate Understanding (UU) exists that can somehow be abstracted from all the data that exists in the world. These abstractions necessarily exist only in the minds of the people who make them. When they are organized into groups, they can become complex mental models of the world that allow us to make accurate predictions about the way the world will behave.

It seems to be a basic human characteristic to project our mental models of the world into the real world as if those models had a separate existence in the external world. People in ancient times did the same thing. Our next step after projecting our mental models into the external world is to insist that those relationships actually have a separate existence in the world. At that point, those relationships wouldn't just *describe* the data in the world, they would be the *source* of that data. Those relationships would no longer simply describe the world, they would *control* the world. This alternate way of viewing relationships separates modern philosophers, scientists, and thinkers (but not the rest of us) from the Bible writers. We would fail to understand these writers if we failed to recognize this crucial difference. The relationships or mental models that the Bible writers project into the world are called angels in the Bible.

You may have a different idea about angels. What you think of angels will depend on many things, but if you follow closely how the Bible writers refer to angels, you may notice that their common characteristic is that they are all messengers. For us, a messenger is someone who carries a message from one place to another. An angel is a more generic concept of a messenger. They are relationships that are energized, and they use their energy to energize other relationships that they control. Energy or spirit is the message that an angel carries. An angel, as a relationship, modifies the energy it receives and delivers that energy to lower-level angels who do the same thing. Eventually the energy is delivered to objects in the world, and those objects move.

Angels of God would be messengers sent from God. Yet some angels are not sent from God. If you bear with me, I think you will find that the way I've defined angels is the larger concept. This definition most closely matches the entirety of angelic behavior in the Bible.

From our perspective, the physical world is outside of us, and it clearly has an existence independent of human existence. Our internal mental world, on the other hand, depends on us for its existence. In addition, we have a compulsion to project our mental models of the world into the world itself. When I say "rock," I've communicated using a word. The word itself is a model, and that model represents another model that exists in my mind and in yours. Other creatures possess models of the world as well. So a world devoid of life should contain no model that we could refer to with the word "rock," and that world devoid of life would contain no model that we could refer to with the word "number." So "rock" and "number" seem to have no existence independent of living things and independent of humanity.

Most highly educated people think that arguments like the one above are mostly settled in our time (or are useless arguments that lead nowhere). But it is fairly clear that many people still believe that the concepts they hold have a separate external existence, even though those models exist only within their minds. If you're one of those people, you might believe that talking about an unwelcome relative might cause him to appear at your door. Or you could believe that picking up a penny that's face down is bad luck. You could even believe that beauty is *not* in the eye of the beholder. The idea that your mental concepts or models or relationships have the ability to affect the external world are common and ancient. So it is not unreasonable to suppose that such a widespread belief was also held by these authors. What might come as a surprise to well-educated people today is that a model of the world based on the idea that relationships and mental models have a separate existence *can* work to effectively model the world. Such a philosophy seems unreasonable to many, but the ultimate question of whether models have an independent existence in a world devoid of humanity and devoid of life is not an unreasonable question to ask, and a positive answer is also not unreasonable.

The fact that understanding is a kind of knowledge implies a deep and mysterious duality between knowledge and understanding. I think the Bible writers were aware of this duality, and with their projection of their mental models onto the external world, they recognized the power of knowledge and understanding as "angels" to control the world. That knowledge and understanding would reach its pinnacle and be concentrated in an Ultimate Meta-knowledge Relationship (UMkR). When we project that UMkR into the external world, it would have *produced* all the data in the world and would also *control* everything in the world. These are the characteristics that they (and we) attribute to a supreme being. So they thought of and spoke of the UMkR as God, and it is clear that they believed that God existed in the external world.

In the New Testament, knowledge and understanding, as God, are separated into God the Father and God the Son.[11] God the Father is the UMkR (that determines all that is knowable in the world) *and* all the data that is knowable in the world. God the Son is the collection of all the meta-knowledge relationships that can be abstracted from the data in the world. All the data is at the bottom, and as abstractions are made from that data, those abstractions can be stacked one upon another until a single all-encompassing abstraction called the UMkR is reached at the top. The ultimate relationship is at the top, and we can think of the next lower level of meta-relationships as being created by the UMkR. These meta-relationships, in turn, create the

---

11 This separation was done in the Old Testament as well, where the Son of God is referred to as "the word of the LORD." In John 1:14 the "word of the LORD" is equated to Jesus, whom we know as the Son of God. When the writer created this equivalence he was implying that the "word of the LORD" from the Old Testament was the Son of God.

relationships that are subordinate to them. This process continues until the relationships that describe and control all of the data in the world are created. God the Father as both the UMkR and all the data in the world would be the Alpha and the Omega (the beginning and the end). God the Son as the collection of all the relationships that exist between the UMkR and the data in the world is all the letters of the alphabet in between.

3.2.1 God the Son and Energy

God the Father is the ultimate meta-knowledge and all that is knowable. God the Son is the means by which that ultimate meta-knowledge is transformed into all that is knowable. God the Father is fixed and unchanging: the UMkR does not change, even though it is the ultimate cause of all change. And the objects in the world do not change of their own accord, though they are affected by change. God the Son is the active principle. When objects in the world move, a relationship, or angel, is responsible for that motion, and that responsibility goes up the relationship pyramid to the UMkR at the top. If a relationship does not cause an object in the world to move, it must forever remain unchanged; it must forever remain at rest. So the deep duality between knowledge and understanding that ultimately makes them the same thing also implies the oneness of God. God is the UMkR, God is all the objects in the world, and God is all the relationships that describe and control everything. God is all one thing, so we can refer to the parts of God interchangeably. Only where the distinction is important does it need to be made.

The process of creation as described in Genesis 1 is the ultimate meta-knowledge (God the Father) transforming itself into all that is knowable (God the Father). All that is knowable is all that exists; it's all the data, it's the world around us. All the meta-meta-meta-[...]-meta-knowledge required to produce the real objects in the world is God the Son. Each level of relationships creates the next lower level until real-world relationships are created. This is what is going on in Genesis 1, where God the Son creates himself over and over again as he creates the world.

Let's look at the details of the relationships that are God the Son. A relationship that describes a collection of data represents some kind of compression of that data. Because a relationship represents a compression of the information contained in the data, there must be fewer relationships as the level of abstraction increases.[12] Relationships can also be grouped into collections where the members are related to one another. We will call such a group of related relationships "models" or "concepts" or other equivalent words. Each relationship or collection of relationships is an angel, cherub, demigod, or some other spiritual entity. The Bible writers held the view that these spiritual entities could control a part of the world, or had the ability to control other relationships that were in control of some part of the world. In their entirety, the collection of all these relationships are God the Son, and they are all energizers of some sort.

Let me provide some Bible verses to show what I mean. From John 1:1–3, we have the following famous lines:

> In the beginning was the Word,
>   and the Word was with God,
>   and the Word was God.
> He was in the beginning with God;
>   all things were made through him,
>   and without him was not anything made that was made.
> In him was life,
>   and the life was the light of men.
> The light shines in the darkness,
>   and the darkness has not overcome it.[ar]

In this passage, God the Son is called the Word, and it is made clear that the Word is God. It's made clear that God the Father is God. It's also made clear that the Word and God are separable (the Word was *with* God), and that idea, together with the stated oneness of God, implies a deep and mystical duality. When creation began, God the Son was the active principle; God the Father was not.

In the famous lines of Genesis 1:1–3, God the Son creates himself in multiple forms—the wind, the voice, and the light:

---

12 The idea that there are fewer relationships is a simplification. Data in a collection can be arranged in so many ways that the number of relationships is *much* larger than the amount of data in the collection. Since any collection of data can create so many relationships, the fact that those relationships ultimately produce the same data makes them equivalent to one another. If we consider equivalent relationships to be the same, the number of relationships is dramatically reduced.

> When God began to create the heaven and earth—the earth
> being unformed and void, with darkness over the surface of
> the deep and a wind from God sweeping over the water—God
> said, "Let there be light"; and there was light.[as]

The sheer density of the information in these three verses makes
them very difficult to understand. God the Father begins to create the
heaven and the earth (or everything that exists, if you will). Out of
him emanates a wind that is the active God the Son. The wind did not
start blowing on this first day, so it seems to have always been
blowing. The blowing wind implies that the creative God the Son has
always been active, and it is he who is actually creating the world.
When God says, "Let there be light," the voice of God (which is the
wind sweeping over the water) is operating to create the light. The
voice of God and the wind in these verses are the same thing, and
they are also the Word of John, Chapter 1. (Put your hand in front of
your mouth and speak. You will feel the wind.) In addition, the light
of Genesis 1 is the "light of men" and the "life" of John 1.

The Bible writers frequently use voice and wind to symbolize
spirit. And a spirit is what energizes a thing and makes it go. This
symbology results from the observation that a wind blowing through
a tree energizes the leaves of the tree. So wind and the sound it
makes, and the sound of a voice and the wind accompanying speech,
were very common analogs that everyone knew. So from the point
of view of the Bible writers, you have a wind inside you that makes
you go. That wind is what keeps you alive. It is your life. Life is about
storing energy, so that energy can be exhausted. All the things that
can use up energy, or life, are called death. That death is also called
darkness. The darkness is said to be "consuming" the light. The idea
that darkness consumes light shows itself in this passage from John
1:5:

> The light shines in the darkness,
> and the darkness has not overcome it.[at]

We know that darkness is the absence of light, and we could
conclude that these authors held the archaic and incorrect view that
darkness was a substance itself. If so, it would explain how they could
believe that darkness could consume light. But just as light is
symbolic in these passages, darkness is as well. So the writers are not

making a statement about darkness itself. They are using darkness to model and to symbolize death.

Other common wind and sound generators are spun into this common symbology. The most common are fire, which generates its own wind (and light), and moving water, which commonly generates sound. Fire associates wind with light, and this association allows us to make sense of the movement from wind, to voice, to light in the first few verses of Genesis 1.

The light of Genesis 1 is the "light of men" and the "life" of John 1, and it overcame the darkness that covered the "surface of the deep [water]." The deep water is symbolic of death. Moving water is "energized" or "living" water; the sound it makes associates it with the wind. Fire associates the wind with the light, and through this association, light is associated with moving water.[13] When moving water flows into a larger body of water, it loses its "life." The larger body of water has "killed" it. The large body of deep water is dark at the bottom, and that association is why the deep water is symbolic of death. When we look up into a night sky, we see the same darkness we would see at the bottom of a large body of deep water. So, by analogy, the darkness above can be viewed as being caused by "waters above." In Genesis 1, the darkness and the deep waters are one and the same.

Examples of this symbology include the story in John 5, where Jesus heals a man who had been ill for 38 years. He, along with many others, waited around a pool for the waters to be "troubled." The first one to enter the energized waters could be healed. However, this man had never made it to the energized waters before someone else did. So Jesus healed him, and in doing so made himself equivalent to those energized waters. Another example is Jesus' teaching that rivers of "living" water would flow from the belly (which means the heart) of anyone who believes in him.

One of the more interesting uses of this symbology is in the ministry of John the Baptist where he baptizes in the flowing water of the Jordan River. The baptism in moving water symbolizes baptism in life itself. This method of baptism could be what distinguishes John's

---

13 In Revelation 15:2 (RSV), John says, "And I saw what appeared to be a sea of glass mingled with fire [...]." This symbolism even more closely ties fire (with its light) with water, giving us a different kind of "living water." John doesn't know that the sea he is looking at is water, since fire mixed with water doesn't make a lot of sense, so he thinks it might be glass.

ministry from the other baptizers of the period. Baptizing in still waters would break that symbolism, but it would have a symbolism of its own. The still waters symbolize death, so baptizing in still waters would represent death and resurrection.

The "word of the Lord" and "the Son of God" are one and the same. John 1:14 makes this clear by equating Jesus, who called himself the Son of God, with the word of the Lord. Here's how we know that an angel is a messenger: If you examine how the Bible refers to the "word of the Lord," you will find that the "word" is not a message, but a messenger. So for instance, in Genesis 15:1, the word of the Lord came to Abraham in a vision and spoke to him. And again, in Genesis 16:9, an angel of the Lord came to Hagar and spoke to her. Both the word of the Lord and angels transmit messages from God to man. The equivalence of the Son of God, the word of the Lord, and the oneness of God allows us to understand that the "word of the Lord" and the "Son of God" are ways to refer to all intermediaries between God and man—all angels. By extension, the word of the Lord and the Son of God are intermediaries between God as the UMkR and the objects in the world.

3.2.2 Faith

Faith is about knowing what is going to happen. It's an important subject in the New Testament. I noted above that given enough understanding of a subject, it is possible in any given situation to have foreknowledge of a future situation, related to the current one, that is just as good as directly observing that future situation. Faith is defined by the writer of Hebrews 11:1:

> Now faith is the assurance of things hoped for, the conviction of things not seen.[au]

This is clearly a reference to foreknowledge, and I think it is a reference to foreknowledge resulting from deep understanding. The deeper the understanding, the further into the future one should be able to predict with high accuracy.[14] *So faith, as defined by the Bible*

---

14 Here, I am using the term foreknowledge as a general term not necessarily limited to relationships that include time as a component. For example, you can have foreknowledge of the next number in the sequence 1,1,2,3,5,8,13.... You can see a major crater impact site at one point of the globe and predict seismic damage at the same point on the opposite side

writers, is a matter of being certain—certain of the understanding and the knowledge that you possess. Certain of the ultimate outcome of all things. You know what is good and what is evil.

Faith is about making *accurate* predictions. Predictions require both knowledge and understanding. The knowledge on which a prediction is based is data about the current situation. The understanding on which a prediction is based allows us to say with high accuracy or with low accuracy that a future situation, related to the current situation, will exist. People who have little faith have little understanding. That lack of understanding means that their predictions are of low accuracy, so they cannot be sure of the future and their place in it.

For example, suppose we had two cars five miles apart on a straight road. The first car is traveling at a constant speed of five miles per hour. The second car is traveling at a constant speed of one mile per hour. At what point will the cars meet?

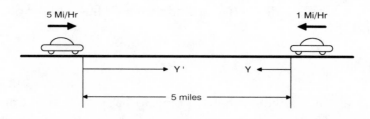

© 2011 Leonard Timmons
Figure 3.1: Foreknowledge. It is possible to know what is going to happen before it does.

This situation contains enough information to allow us to predict where the cars will meet. The knowledge that we can apply to this prediction is provided above. The *understanding* that we can bring to bear to make our prediction can vary greatly, however. For instance, most people will be able to make the low-accuracy prediction that the cars will meet closer to the starting position of the slower car. Others will make their prediction by using algebra to say that when their positions are equal, they are at a position where Y and Y' are equal. During the travel of the slower car, its position is

Equation 1: $$Y = 1 \times T$$

of the globe.

where T is the amount of time that has passed since both cars were at their starting positions and the number one (1) is the number of miles per hour the car is traveling. During the travel of the faster car, its position is

Equation 2: $Y' = 5 \times T$

where Y' is the distance from its starting point, T is the time that's passed, and the number five (5) tells us it is moving at five miles per hour. But we note that

Equation 3: $Y' = 5 - Y$

which tells us that we're on a fixed, straight track, and the location of the faster car can be measured from the starting point of the slower car. So the faster car's position will be

Equation 4: $5 - Y = 5 \times T$

or

Equation 5: $5 - (5 \times T) = Y.$

In equations 1 and 5, the value of Y is the same when the cars meet, so we can find the value of T:

Equation 6: $5 - (5 \times T) = 1 \times T$

and the cars meet when

Equation 7: $T = 5/6$ hour

(that is, they will meet 50 minutes after they start), and

Equation 8: $Y = 5/6$ mile

from the slower car's starting position (and $Y' = 4\text{-}1/6$ mile from the faster car's starting position). Clearly this is a much more accurate prediction. We can make the prediction and be certain of the outcome. We have faith in (that is, we are certain of) the outcome, though we have not seen the cars meet. This high-accuracy prediction is so accurate, it is almost as good as seeing the cars meet on that road.

When we make a prediction, whether we are right or wrong is determined by the accuracy we need. It is easy to make predictions and have faith in the outcome if you never have to meet any test for accuracy. But we can't really call this faith. Faith has magnitude. Great faith involves predictions that have very close tolerances. In our example above, it requires little faith to predict that the cars will meet closer to the starting position of the slower car. Most people can make that prediction and be certain of the outcome. It requires an understanding of algebra to make the prediction that the cars will meet 5/6 mile from the starting point of the slower car. The tolerances associated with the algebraic prediction (great faith) are much, much tighter. And the understanding you must have to make the prediction using algebra is much, much greater.

We should look at predictions, and the faith involved in making them, in terms of an archer who is trying to hit a target. When an archer is two feet from a two-foot-diameter target, the amount of skill (that is, faith) required to hit the target is low. However, an archer who is 20 feet away from a two-foot target needs some skill to hit the target. An archer who is 200 feet from a two-foot target must have tremendous skill to hit the target.

If there is never a judgment, then faith is never tested. Without a judgment, you will never find out if your prediction was right or wrong. In the Bible, righteousness is about being right. It is closely associated with faith and understanding and knowing what is going to be. In Genesis 15:4–6, the Lord speaks to Abram:

> And behold, the word of the LORD came to him,
>> "This man shall not be your heir; your own son shall be your heir."
> And he brought him outside and said,
>> "Look toward heaven, and number the stars, if you are able to number them."
> Then he said to him,
>> "So shall your descendants be."
> And he believed the LORD; and [the LORD] reckoned it to him as righteousness.[av]

Abram (Abraham) would not be around to see his descendants. But he understood that the Lord had the power to accomplish his promise. He saw deeply into a distant future where his descendants could be counted only with great difficulty. We know that Abram's prediction was right, and that rightness made Abram righteous. He showed great faith.

3.3 More Insight

Insight is the means by which understanding is produced. Insight is all about finding relationships in the data, and that data is knowledge. Even with only a little knowledge, great insight can lead to great understanding that produces foreknowledge that is as certain as knowledge gained by direct observation.[15]

Since insight produces understanding and understanding is a type of knowledge, and since direct observation also produces knowledge, it's easy to think of insight as a type of indirect observation. If we extend this analogy, we have normal sight with our physical eyes, and we have insight with our "spiritual eyes." Spiritual eyes imply a spiritual body, and that body would be the soul. If we consider this analogy to be more than an analogy, if we think of it as actually representing the external world (as the Bible writers did), we could conclude that by demonstrating that you have insight, you produce evidence that you have a soul.

Since it is so easy to think of insight as a type of eyesight, a great symbol for insight is the eye. And since the knowledge produced by insight is data that is the result of direct observation by an "inner eye," eyes within and without would accurately symbolize a person with insight. The more eyes, the more insight. With this information, the symbology of Revelation 4:8 becomes clearer:

> And the four living creatures,
> each of them with six wings,
> are full of eyes all round and within [...][aw]

The living creatures represent a person or spirit of observation.

---

15 A prediction requires knowledge of the current situation and operates to give us the details of a future situation. As a person's understanding increases without limit, his power of prediction also increases without limit. When that person achieves Ultimate Understanding, he does not need any situational knowledge to make a prediction about all future situations, and those predictions are made without error (more on this in Section 7.5.1).

They have eyes on the outside to gather lots of high-quality data about the world. And they have eyes on the inside (insight) to generate understanding from that data.

Today, our common symbol for insight is a light bulb shining over someone's head. We frequently use this symbol to indicate that a person has suddenly understood something that was very difficult. This light symbology is ancient, but the form has changed due to changing technology. In the New Testament, Jesus promises to send a comforter, his Holy Spirit, after his death. This Spirit shows itself on the day of Pentecost with a great rush of wind and a tongue of fire that rests on everyone's heads. The symbology of the wind is covered above. But the tongue of fire above the head is the same symbol as our light bulb symbol for insight.[16] And the promised comforter is the Spirit of Insight, which we call the Holy Spirit. The reason we would expect to be comforted by such a spirit is that the insight it brings allows us to have greater understanding so that we can make predictions of much greater accuracy, and that accuracy allows us to be more sure, even certain, of the future and our place in it.[17]

When light penetrates darkness, the light makes it possible to see the things that were hidden by the darkness. The light itself is not insight. The light enables insight. Insight is the act of observation on deeper and deeper levels. Insight is seeing God.

The Bible was written to test for insight. It is a book intended to confuse, to darken. When you have insight, you have the ability to emit a spiritual "light," and that light shines through the darkness of the book, allowing you to observe what is hidden within it. Today we would say that a person with these kinds of talents is "brilliant." The Bible was not written to enlighten as we would expect to be enlightened. It was written so that its readers could provide evidence of *their* enlightenment—evidence of their souls.

The Bible also seems to have confusing information about seeing God. God the Father as the UMkR is at the top of what could be an infinitely high pyramid. Yet God the Father as the material world is immediately at hand and within reach. We can all experience the material world, and almost all of us can understand simple relationships in that world. Understanding those relationships is

---

16 This symbology is even more ancient. When Moses encountered the bush that burned but was not consumed, he was also encountering a massive insight about how to free his kinsmen, the children of Israel, from Egypt.

17 What insight did Jesus' followers have on Pentecost? I do not know.

seeing a small part of God. Others can understand the relationships that control and describe much more of the world. Those who understand those relationships can see much more of God. The Bible writers let us know that it is almost impossible for any person to experience the God who exists at the top of this relationship pyramid. You see evidence of all of this in passages where it is said that no one has seen God at any time. Then in other passages, it is made clear that God *has* been seen. At other times, the Face of God cannot be seen, but the Body of God can. Consider these passages from the Revised Standard Version of the Bible:

> Genesis 32:30: So Jacob called the name of the place Peni'el, saying, "For I have seen God face to face, and yet my life is preserved."

> Exodus 33:7–11: Now Moses used to take the tent and pitch it outside the camp, far off from the camp; and he called it the tent of meeting. And every one who sought the LORD would go out to the tent of meeting, which was outside the camp. Whenever Moses went out to the tent, all the people rose up, and every man stood at his tent door, and looked after Moses, until he had gone into the tent. When Moses entered the tent, the pillar of cloud would descend and stand at the door of the tent, and the LORD would speak with Moses. And when all the people saw the pillar of cloud standing at the door of the tent, all the people would rise up and worship, every man at his tent door. Thus the LORD used to speak to Moses face to face, as a man speaks to his friend. When Moses turned again into the camp, his servant Joshua the son of Nun, a young man, did not depart from the tent.

> Exodus 33:18–23: Moses said, "I pray thee, show me thy glory." And he said, "I will make all my goodness pass before you, and will proclaim before you my name 'The LORD'; and I will be gracious to whom I will be gracious, and will show mercy on whom I will show mercy. But," he said, "you cannot see my face; for man shall not see me and live." And the LORD said, "Behold, there is a place by me where you shall stand upon the rock; and while my glory passes by I will put you in a cleft of the rock, and I will cover you with my hand until I

have passed by; then I will take away my hand, and you shall see my back; but my face shall not be seen."

Judges 13:17–22: And Mano'ah said to the angel of the LORD, "What is your name, so that, when your words come true, we may honour you?" And the angel of the LORD said to him, "Why do you ask my name, seeing it is wonderful?" So Mano'ah took the kid with the cereal offering, and offered it upon the rock to the LORD, to him who works wonders. And when the flame went up toward heaven from the altar, the angel of the LORD ascended in the flame of the altar while Mano'ah and his wife looked on; and they fell on their faces to the ground. The angel of the LORD appeared no more to Mano'ah and to his wife. Then Mano'ah knew that he was the angel of the LORD. And Mano'ah said to his wife, "We shall surely die, for we have seen God."

1 John 1:18: No one has ever seen God; the only Son, who is in the bosom of the Father, he has made him known.

1 John 4:12: No man has ever seen God; if we love one another, God abides in us and his love is perfected in us.

In this collection of passages, we can glean that the UMkR is the Face of God. The Body of God is what we have been calling God the Son, or the myriad relationships that exist between the UMkR and the objects that exist in the world. To see God or an angel is to have an insight. The insight that allows us to see the Face of God (the UMkR) is the ultimate insight. Such an insight is presumed to result in the loss of one's life. I think the authors would like us to believe that this is the insight that Enoch had when he is said to have "walked with God," and God took him.

In the first scripture cited above, Jacob clearly deals with an angel whom he thinks is God the Father (the UMkR). In the second scripture, where Moses also encounters an angel, the writer talks of the angel as if it were God the Father, but cannot be, since Moses could speak with the angel face to face. In the third scripture, it is clearer that Moses is dealing with God himself, both the Father and Son. And it becomes apparent that though it might be possible for Moses to see the Face of God, he would not live through it, so God makes sure that Moses does not see his face. In the fourth scripture, Mano'ah and his wife make the same mistake by confusing an angel with the Face of God. In the last two scriptures, John makes it clear that God the Father (as the UMkR) has never been seen. Never. Only God the Son has been seen (via an angel, for instance), and only through him can God the Father be known. John is saying that God the Father, in the form of the UMkR, exists as an extrapolation to the single all-encompassing rule from which all the relationships that make up God the Son are made. (Seeing the "back" of God is another way to say the same thing.) The oneness of God allows these speakers to consider any individual angel or collection of angels to be "God the Son" or "God the Father" without being completely wrong.

3.4 Wisdom

Wisdom is *knowing what to do* with the knowledge and understanding that you have. Wisdom, like understanding, is a type of meta-knowledge—it is knowledge of knowledge. Wisdom is knowing how knowledge and understanding impact our lives.

What should a person do with what he knows? And what shouldn't a person do? The second question is answered definitively at the very beginning of Proverbs:

> The proverbs of Solomon, son of David, king of Israel:
>
> That men may know wisdom and instruction, understand words of insight, receive instruction in wise dealing, righteousness, justice, and equity; that prudence may be given to the simple, knowledge and discretion to the youth—the wise man also may hear and increase in learning, and the man of understanding acquire skill, to understand a proverb and a figure, the words of the wise and their riddles.
>
> The fear of the LORD is the beginning of knowledge; fools despise wisdom and instruction.
>
> Hear, my son, your father's instruction, and reject not your mother's teaching; for they are a fair garland for your head, and pendants for your neck.
>
> My son, if sinners entice you, do not consent.

If they say, "Come with us, let us lie in wait for blood, let us wantonly ambush the innocent; like Sheol let us swallow them alive and whole, like those who go down to the Pit; we shall find all precious goods, we shall fill our houses with spoil; throw in your lot among us, we will all have one purse"—my son, do not walk in the way with them, hold back your foot from their paths; for their feet run to evil, and they make haste to shed blood.

For in vain is a net spread[18] in the sight of any bird; but these men lie in wait for their own blood, they set an ambush for their own lives.

Such are the ways of all who get gain by violence; it takes away the life of its possessors.[ax]

The phrase "the fear of the Lord" requires some explanation, since it is used repeatedly in the Old Testament. This is not a direct reference to being afraid of God because God is "out to get" anyone who does something he doesn't like. The best way to think of this kind of fear is to imagine that you are lost in a huge power plant that is generating 600 megawatts of power—enough to power an entire city. In that situation, you must respect the power of the plant. The power plant is not "out to get you," it is trying to provide power. While you're in the plant, you could do something foolish that could result in your immediate and spectacular death. Any reasonable person in that situation would be afraid. To keep from seriously injuring or killing yourself, you must learn as much about the plant as quickly as you can. You have a real need to know where you are, what is around you, and what it takes to get out of the plant alive. That fear is the beginning of knowledge, and that knowledge isn't a luxury, it is a necessity.

The clear implication of this explanation is that we should not think of God as a person, but rather as the physical world around us. That world is a very powerful thing, and its processes proceed without regard for our presence. We are each lost in that world, and we're trying to get out of it "alive." Our lives can be lengthened or shortened by our actions. And our actions are determined by how much we know about the world. Each of us has an unmistakable

18 The net is a trap.

obligation to know as much as we can about the world. We can get that knowledge by being taught, but our knowledge of the world is so incomplete that each of us also needs to have the ability to figure things out for ourselves. Not only do we need to be willing to learn, we need to have a talent for insight.

So the fear of the Lord is real fear, and that fear is the beginning of knowledge. Not knowing is being in darkness. Not knowing is sin.

These first verses of Proverbs tell the initiate what *not* to do with the knowledge and understanding that he possesses, and by extension they tell us what not to do as well. It concludes that destroying yourself with your knowledge and understanding is foolish. Foolishness is evil, and the failure to possess knowledge and understanding (ignorance) is also evil. Conversely, using your knowledge and understanding to make yourself live is good.

To the Bible writers, *using your knowledge and understanding to make yourself live is the ultimate and the only good.* Failure to use your knowledge and understanding to make yourself live is the only evil. And wisdom is all about making yourself live. That is, wisdom is all about being good. Foolishness is all about not making yourself live, or actively destroying yourself, or embracing your own destruction—it is all about being sinful, or evil, or wicked, respectively.

The ultimate good is eternal survival, and eternal survival is everlasting life. If everlasting life can be achieved, it would have to be achieved by knowing in every survival situation exactly what you should do to ensure your survival. But how could you possibly know what to do in every situation? Each situation is unique, even though it might be similar to past situations. So how could you effectively deal with unique situations that can affect your survival adversely? Your only hope of doing this would be to develop an extremely deep understanding of the world. With that kind of understanding, you could make *accurate predictions* over arbitrarily long timespans. If you should achieve a perfect understanding of the world, you should be able to accomplish what would otherwise be an impossible task. Unfortunately, to accomplish the impossible task of predicting everything that would impact your survival, you would have to obtain a perfect understanding of the world.

In reality, you don't understand the world perfectly, but you would be wise to try to understand it as best you could. So to survive as long as possible, you would need to do the right thing and not the wrong thing in any survival situation. In order to do that, you must gain as much understanding of the world as possible so that you can make accurate predictions about what will happen. Those accurate predictions are your faith—the more accurate, the more faithful. So it is your *faith* that gives you the ability to do the wise thing and not the foolish thing in any survival situation. Or said another way, the magnitude of your faith determines whether it is more or less likely that you will live forever. Yet another way to say the same thing is, *the magnitude of your faith determines how good you can be.*

The more faith we have, the more accurate our predictions about what the future will bring will be. So faith allows us to see threats to our lives before they materialize and to act to eliminate those threats. If I act to eliminate a threat to my life, that would be a good act, because it helps to ensure my survival. The failure to act would be evil. If I could avoid doing evil altogether, my survival would be forever guaranteed.

So it is the acquisition of faith that must be our primary goal in life if we want to be as good as we can be or live as long as we can. When we acquire faith for ourselves, we improve our chances of survival. Each act we take to improve our chances of survival is an act of love. Each act we take to improve someone else's chance of survival is also an act of love. One of the best ways to help someone else improve their chances of survival is to help them acquire faith. Faith requires understanding, and great faith—the faith to be very, very good—requires great understanding. Great understanding must be acquired somehow, and that understanding is acquired through great insights. When you know what to do with what you know and understand, you use that knowledge and understanding to extend your life and the lives of others. As your knowledge, understanding, insight, and wisdom increase without limit, your lifetime must also increase without limit, and you would live forever.

3.4.1 The Soul of Everlasting Life

Everlasting life is a concept that is a clear problem for the biological creatures that we are. Ultimately we all fail to continue living. In fact, everlasting life for a person suggests that some part of that person, the thing that makes him a person, can somehow maintain its integrity and persist indefinitely. In the Bible, the thing that can survive indefinitely is called a person's soul.

A good analogy of a soul is the software you can run on your computer. When a software program is running, it is energized and "living." A "software soul" is "dead" when it is not running on a computer, but as soon as you start it up again, it lives again. A computer program is a set of instructions to recreate a process. When those instructions are fed into a computer and run, the process it describes comes alive. Every process that can be reduced to a set of instructions has a soul, and that soul is those instructions.[19] So there are souls everywhere. Each cell in your body has a soul. The cells that make up your liver are a part of a larger process, and that process has a soul as well. The "soul of a liver" is recorded in your DNA, and those instructions allow that soul to live again in your children. The collection of organs in your body is a part of a larger process called the human body, and that body has a soul as well.

You might notice here that an energized program bears a more-than-passing similarity to an angel. It is, in fact, equivalent to an angel. The more general concept is that an angel takes inputs and transforms them to produce outputs. In this more general definition, the inputs could be anything, not just spirit or energy, and the outputs would be *related* to the input. That relationship would define the angel (that is, the relationship would tell us what kind of angel we are dealing with). When we think of this in terms of the UMkR, it is the collection of all the angels (also known as the word of the Lord or the Son of God) that transforms the spirit of the UMkR into the motion of objects in the world.

You are a collection of millions of processes; you are a collection of millions of souls. Some of those processes are common to all human beings, and the instructions that reproduce those processes are the "soul of a human being." Though that soul is "you," it is also everyone else. Other processes are not common to all human beings. Some of those processes are common to groups of humans, while others are unique to individuals. The processes that are unique to you create your individual soul. If we had the technology to take those instructions and implement them, "you" would suddenly appear. If we were to save those instructions, we could bring "you" to life at will.

Souls are complicated by the fact that the underlying processes can change over time. Each change produces a new process with a new soul, though the new soul is related to the original. I've changed since

---

19 Only processes that cannot be reduced to a set of instructions would not have souls.

I was a child, and I suspect you've changed as well. I used to have the soul of a child; now I have the soul of an adult. The processes that are unique to me, that make my soul unique, are changing, and what is "me" today may not be "me" five years from now.

Processes that are composed of other, smaller processes complicate matters further. A processor that can give life to more than one process decouples the process from any particular arrangement of matter. For instance, a computer can run a program that controls a production line, or it could run a program that helps design bridges. A flexible underlying processor can allow all kinds of processes to run. Processes that have never existed before can be created and given "life." The limit of this kind of flexibility for living creatures is the brain and the mind that executes in that brain. That brain allows the highly flexible living thing that has evolved such a brain to survive better. The mind expressed from that brain has the task of helping the living processor on which it runs to remain alive. The wise mind is better than the foolish mind at helping its underlying body and brain remain alive. Our minds contain knowledge, understanding, and wisdom. When our minds have insights, those insights help increase that understanding. A mind is characterized by the instructions and data that it contains, so a mind has a soul contained within the brain on which it runs.

To call someone foolish is to say that their minds are not performing their primary function of keeping the underlying body alive. Such a mind is no mind. A soul that contains no instructions on how to to keep the underlying body alive is no soul. So when you prove that your mind can have insights, you give evidence that your mind can work to keep the underlying body alive—you provide evidence that you have a soul, a real soul.

When it is running, a soul creates a mind. And that mind is able to change the capabilities of the underlying soul as it gains more knowledge and understanding of the world. I could train to be a plumber, and once I'd finished my training, my soul would contain the soul of a plumber. More importantly, my mind could change the plumbing processes and change what it means to have the soul of a plumber. Each person's mind can seek out process instructions from other minds to add to its soul, its collection of process instructions. A person's mind can recognize its own deficiencies and seek to correct them by adding new processes to its collection. That person's soul would change as a result. This is the experience of consciousness.

Consciousness is a mind thinking about the soul that produces it.

Our individual minds are not advanced enough to keep our individual bodies alive forever. Every process intended to keep a single biological creature alive and unchanged (as far as we know) ultimately fails. Because individuals die, reproduction allows the processes executing on that individual to outlive the individual. Because collections of individuals can become extinct, evolution allows the individuals to change over time into new individuals who are, as a collection, more able to deal with threats to their collective survival. The struggle to survive goes on. Is there an individual who cannot be killed? Someone who will live forever? Is there a family that cannot be killed, even though its members die? Is life on earth incapable of being eradicated, even in the face of the extinction of families? The soul of any of these life processes could be said to be in the process of living forever. We would have to characterize any such process, any such soul, as being extremely wise. I'm pretty sure that my body will not live forever.[20] It is not yet clear that humanity is capable of living forever. But life on earth has been here for billions of years, so an argument can be made that it is in the process of living forever. Can it survive the destruction of the earth? That's less likely.

Souls are collections of data, just like relationships are, so they can be moved from place to place just like any other data. That collection of data is knowledge and understanding. The purpose of the soul is to keep the body alive, so from the point of view of the Bible authors, your soul is the collection of relationships that describe and control you as an individual—your body and your mind. We often presume that we can separate the soul in your brain (that produces your mind) from the soul of your body, and the soul of your brain could still produce "you." It certainly may become possible that some future technology will allow your mind or your brain to be transferred to another body and "you" will appear.[21] When defined in this way, the soul that produces your mind is ultimately raw data, and that data

---

20 My body has evolved to live, reproduce, and get out of the way of the next generation. That is the soul of its life process. It would take a radical change in its programming to make it behave differently. I don't have the technology to make that happen.

21 We know that our sensory organs do a lot of preprocessing of data before our brains get the "tokenized" information that those sensory organs produce. If we consider the locus of the mind to be in the brain, then separating the mind from the preprocessing of its sensory organs could leave it disconnected from the world. If we consider the mind to be distributed throughout the body, then separating the mind from the body involves replicating some or all of our sensory organ preprocessing for that disconnected mind.

can be stored in a data storage system. A data storage system that each of us is familiar with is our own memory. So it is conceivable that someone with a very, very good memory could store someone else's soul. That person might not be able to give life to the stored soul by energizing it and becoming the other person, but we expect that something like that could possibly happen. Since each of us can relive our memories, what's to say that we couldn't "slip into" our mental model of another person and allow that person to live again?[ay]

We've been talking about everlasting life, but I have failed to say precisely what life is. Unfortunately, defining life is beyond the scope of this book. Please see my upcoming book, *Life and Life on Earth,* for a more thorough discussion of the subject. Certain basic characteristics of living things are presumed as obvious. Living things seek to remain alive; living things contain energy; and there is a difference between death (being de-energized) and destruction (having one's characteristics changed to an extent that what used to be "you" cannot recover and live again). Successful living things in complex environments need to know what's about to happen (they need to have faith). That knowledge allows them to act in ways that prevent threats to their lives from succeeding in taking away that life.

One of the basic discoveries of modern science is that probabilistic behavior is fundamental to the way the world works. So even though it may be possible to predict the gross behavior of the universe into the indefinite future, it is not possible to predict that behavior with detail. The problem of prediction is huge because the universe is so incredibly complex. So the very idea that something could continue to live into an indefinite future by predicting in detail every threat to its life process and countering each threat is an impossibly tall order. The Bible authors were very aware of the randomness present in the world, and they equated that randomness with darkness. They also understood that the darkness of randomness was a fundamental part of their world. But unlike us, they thought that the randomness/darkness could be evicted from parts, if not all, of the world. Once the darkness of randomness had been evicted from a region of the world, everlasting life would be possible there.

When the darkness of randomness was removed from a certain region of the world, the structure of the world would have to change in that region. (From our perspective, that region of the world would have to become deterministic and its future totally predictable. Or time would have to stop in that region of space, and it would have no future. Or it could be that modern science is wrong and probabilistic behavior is not fundamental to the way the world works; the world is now predictable in principle, but we don't understand the world well enough to make those predictions.) These authors did not see a world that was predictable, so they predicted that the structure of the world would, in fact, change at some future date. It would change so that everlasting life would be possible in that new world. So they spoke of that new world and the everlasting life that was possible there in an inferential manner. They could talk about everlasting life as though it existed, with the knowledge and understanding that it could not actually exist in their world. But everlasting life *did exist* as the best model of the ideal living thing that they could construct. And everlasting life *didn't exist* because everything they were familiar with eventually died. It both did and did not exist at the same time. That new world would be a region where God actually imposed himself on the world we know. That imposition would be known as the Day of the Lord.

3.5 Prophecy

For many of us, the prophets of the Bible are larger-than-life people who were chosen specifically by God to impact the history of his people Israel. Their very existence was miraculous. But the Bible also tells us about a system of education that trained prophets. Elijah and Elisha traveled to a number of cities where they visited "the sons of the prophets."[22] These students knew of Elijah's expected imminent ascension into heaven. After the event, they went out to search, cluelessly, for Elijah's body. It really becomes clear that they were students.

When Elijah and Elisha visited each of the schools, they gave us a glimpse into how this system of education worked. I have no idea how the students knew that Elijah was to be transported directly into heaven, and I don't have a clue how Elisha knew. But the association between these various schools was clear. And describing the students as "the sons of the prophets"[22] tells us clearly that they were training to be prophets like Elijah and Elisha, two of the greatest prophets in history.

---

22 The Bible writers frequently use the phrase "the sons of" to mean "the successors to" or, equivalently, "the trainees of."

But you weren't a failure if you did not become a great prophet. In ancient Israel, the prophets were the repositories of reliability. They were not only the source of the most reliable predictions, they were also teachers, advisers to kings, and the conscience of the nation. Some will disagree that the prophets were reliable predictors of the future. But everyone must agree that the writers of the Bible wanted us to believe they were. And prophecy is all about prediction, of course. The ability to predict is what makes a prophet a great teacher and a great adviser, and that ability forces a prophet to be the conscience of a people.

The essence of being a prophet is the ability to make predictions. Since we all make predictions every day, we are all prophets to some minor extent. A prophet wants to make faithful (accurate) predictions, so he must have faith. To determine whether a prediction is accurate or inaccurate requires that a prediction be *tested*. Either the prophet must test the prediction himself, or those who receive the prediction must test it to see if it is a faithful one. So faith and the magnitude of one's faith is what makes a prophet or a great prophet. *A prophet is a person of faith who has demonstrated that faith.* A prophet is a person who makes predictions that are reliable. A prophet is a person who makes predictions that withstand testing.

Is this what you thought prophecy was about? Maybe not, but this is the way the Bible describes it. In Deuteronomy 18:18–22, Moses tells us what God said to him about real prophecy:

> "And the LORD said to me, '[...] I will raise up for them a prophet like you from among their brethren; and I will put my words in his mouth, and he shall speak to them all that I command him. And whoever will not give heed to my words which he shall speak in my name, I myself will require it of him. But the prophet who presumes to speak a word in my name which I have not commanded him to speak, or who speaks in the name of other gods, that same prophet shall die.'
>
> And if you say in your heart, 'How may we know the word which the LORD has not spoken?'—when a prophet speaks in the name of the LORD, if the word does not come to pass or come true, that is a word which the LORD has not spoken; the prophet has spoken it presumptuously, you need not be afraid of him."[ba]

We can see from these verses that the writer is telling us that true prophecy is about making accurate and reliable predictions, and those who make inaccurate, unreliable predictions are not prophets (of the Lord). The prophet who speaks without understanding, who speaks without faith (accuracy), destroys himself and those who listen to him.

The important thing to recognize in this scripture is that the concept of testability is actually *fundamental* to the system of education that produced the Bible. Predictions (and the relationships that produce them) that can never be falsified are specifically excluded from the arena of prophecy. A prophet *must* be able to make reliable short-term predictions to be considered a prophet. Those of us who listen to him can test those predictions by checking to see how accurate (faithful) they are. Then we can make our own estimates about whether that prophet's long-term predictions are likely to also be accurate (faithful). In fact, the scripture insists that *the individual hearing a prophecy is responsible for knowing the truth when he hears it*, so he is directly responsible for knowing whether what the prophet says is actually true. So the prediction that a prophet of Moses' caliber would arise is a prediction we could trust if we knew that the prophet making the prediction had been highly accurate and reliable in the past.

When described in this way, prophecy sounds a lot like modern science. In fact, it is a lot like modern science. Everyone eventually needs reliable predictions, and cultures develop mechanisms to produce those reliable predictions. In modern society, we've developed a system of making reliable predictions that we call science. Science is different from prophecy in that one of the important components of science is repeatability. Science concerns itself with the repeatable phenomena that occur in our world. Things that happen once and once only are not addressable with the scientific method. In particular, historical events cannot be addressed with the scientific method. Prophets, on the other hand, can and do address historical events. Science is prophecy, but prophecy is not necessarily science.

Science can answer questions like, "If I build this building in such-and-such a way, will it withstand a magnitude-seven earthquake?" That's clearly an important question, and the answer needs to be reliable in order to prevent the deaths of the people living in the building. Science has the power to answer that question, and with that kind of reputation, lots of people want to label themselves with the word. That desire moves us from the hard sciences, which use the scientific method every day, to the soft sciences, which use some modified version of the scientific method.

Other disciplines use repeatability in the sense that "most people agree," which means that repeated agreement is some kind of "repeatability." The hard sciences are topics like physics, astronomy, or geology. Hard sciences are known for repeatedly manipulating things to determine how they will respond. The softer sciences are topics like sociology, archeology, and political science. Manipulation of social objects or political objects, which could be a person or persons, is either difficult, not ethical, or affects the object to an extent that the result of any measurement cannot be relied on. In a discipline like history, manipulating historical events is impossible, but the interested parties can repeatedly agree or disagree that an event occurred as it was reported.

Distinctions like the ones we've made above did not exist in ancient times. Philosophy, religion, science, engineering—everything was all mixed together. Practicality and usefulness were the most important things, and they are more important in engineering than in any other field that we know today. These authors were engineers in spirit, but they were philosophers at heart with scientific souls.

Prophecy includes every possible kind of prediction. Science is certainly not geared toward making historical predictions. But historical predictions can be made. In fact, our minds have evolved to make exactly that kind of prediction. When we want to know what another person will do in a particular situation they have never encountered before, we turn to our mental model of that person and run it through a mental model of the situation to see what happens. That prediction cannot be a scientific one, but it can be accurate enough to save your life, especially if your life is being threatened by another person and you have to determine whether that threat is real.

So what would you teach in a school devoted to producing prophets? Testability. We don't know what went on in these schools, but in the scripture above we see the students verify that they could not find Elijah's body. We can expect that the students were taught that one should determine whether what is reported to be an insight actually *is* an insight. They would teach that one should determine whether what is supposed to be faith is actually accurate and has the accuracy that the task at hand requires. They would teach that you should know that "what you think you know" is actually knowledge. They would teach that your actions are responsible for whether good or bad things happen to you, and that wise actions benefit you and everyone around you. They would teach that the ultimate outcome of being wise is everlasting life and that it is possible to be so incredibly wise you could walk with God like Enoch did, and God would take you, just like he took Elijah.

3.6 What the Bible Is

This chapter is necessary because the differences between us and those who wrote the Bible are enormous. We cannot understand their writings without understanding their worldview. Though this chapter presents my view of their view of the world, I have developed this view based on my reading of the Bible and other research over my lifetime. So the theology I have presented in this chapter is what led me to understand much of what I have understood of the Bible. It is what allowed me to understand that the genealogy of Adam was a puzzle, and that the stories associated with it were riddles. To explain the remaining puzzles and riddles of Genesis 1 to 11, I will need to refer to the philosophy/theology of this chapter.

It is hard to accept much of what I've related above. I know. Angels aren't what you thought they were. Souls aren't what you thought they were. A mind is a living soul—not a definition that most people would readily accept. Faith in the Bible is not religious faith, but is, in fact, the more common definition of something that can be relied upon to be true. And faithfulness does not mean blindness but being reliably accurate. God is the world around us and the relationships that energize that world. And the reason we are alive is to make ourselves live, if not forever, at least for a very, very long time. In what follows, I will show you that this is the way that these authors looked at the world.

So let's summarize. This philosophy/theology begins with the following symbology: wind, voice, sound, light, fire, life, and moving water. Each of these symbolizes spirit. Today we would refer to the common principle underlying these symbols as energy, so energy and spirit are the same thing. Next in this theology is the act of observation. Observation produces data, and that data can be used to produce understanding within an observer. That understanding is also data but is more abstract than the original data, and this duality is deep and mystical. The understanding we produce through insight is abstract data that can itself be understood more deeply, and that new understanding is more abstract than the original data. We can repeat this process indefinitely and possibly produce an ultimate abstraction of all the data that exists in the universe. That ultimate abstraction is the Ultimate Meta-knowledge Relationship (UMkR) or Ultimate Understanding (UU). The UMkR and all the relationships that exist, including the physical world that is being observed, are God. God the Father is the UMkR and the physical world taken together. The relationships that "flow" from the UMkR are what we call the Son of God. The Son of God can be divided into other relationships that represent heavenly beings: cherubim, angels, and other heavenly creatures. These beings energize the objects in the world.

Within each observer is a process of abstracting data. That process is what we call insight. Insight is not different in kind from the process of observation. The symbology for insight is the "inner eye," and since insight works on data and on understanding to reveal relationships, insight is seeing God. With insight you can see an angel. And that means you have understood a relationship that both describes and controls the world. The insight that produces Ultimate Understanding in an observer transforms that observer into a different kind of being, and we see that transformation as that person disappearing from our world. The difficulty of achieving this level of insight is told as our being able to "see" the "body" of God, but not being able to "see" his "face." The insight that operates within an observer to allow him to understand the data that he collects is a mysterious process. The process is facilitated by a "Spirit of Insight" called the Holy Spirit or the Comforter. Helping a person gain insight is an act of love.

With insight and the ability to understand comes the ability to predict. The more insight and understanding you have, the more you can predict. Highly accurate predictions over time and space are equivalent to observed data. The ability to make accurate predictions is faith. Observers with great faith can predict their ultimate destiny, and from those predictions they can determine what to do moment by moment to bring their destiny to pass. Every act of the observer would be a wise act. The theory is that with Ultimate Understanding, you could predict into eternity with perfect accuracy. However, since such a prediction can only be verified at eternity, and since we can't directly experience what will exist at eternity, the prediction cannot be verified (unless the structure of the universe changes). Therefore, the "Ultimate Understanding" that produced the prediction is not certain to be Ultimate Understanding. This Ultimate Understanding is the "face of God," and by this reasoning we cannot possibly experience it. The Bible writers thought that we could "see" the face of God. When God revealed his "face" in a certain place, the structure of the world would change at that place, and everlasting life would be possible there. To these writers, God revealed his "face" to Enoch, and Enoch disappeared as a result.

The practical issues involved with everyday living are also important. First, you have to determine where you want to be at eternity, and then you have to make the decisions that are required to get you there. At each moment, you must make the correct choices that will take you to the place you want to be. The first thing a practical person recognizes is that he cannot possibly do this without error. His only hope is to try to reduce the number of errors he makes at each moment of his life to zero. Our practical person must have Ultimate Understanding to accomplish such a feat. Getting that Ultimate Understanding requires that he obtain insight. Obtaining that insight for oneself is an act of self-love. Helping someone else obtain that insight is also an act of love. To help others obtain insight, you must create a system of education based on finding those who have the "Spirit of Insight," or the Holy Spirit. That system must be designed to help its students gain more insight.

The authors of the Bible or their ancestors created a system of education based on love. That system ultimately produced a book of riddles to help the initiate develop his talent for insight. That book is the Bible. The Bible is a love letter from our ancestors. The student of the Bible can use it as a tool to develop great insight and great understanding. That great understanding gives him the hope of acquiring Ultimate Understanding. If he should gain Ultimate Understanding, we would expect him to disappear from earth like Enoch did and to live forever if forever exists.

The other side of this whole philosophy/theology begins with the symbols of darkness and still water, each of which represent death. Today we would refer to these as energy sinks. An energy sink is a thing that drains energy from something else. If you are alive and something effectively drains all of your energy, then you will die. An effective energy sink is death.

Next on this side of the theological divide is the failure to observe. Observation produces data, and that data can be used to produce understanding within an observer. The failure to effectively observe is exactly the same as collecting poor-quality data. The poor-quality data can produce understanding that is unreliable. Data can be of poor quality for two reasons. It might have been collected with low accuracy because of a failure to effectively observe, or the data could result from accurately observing a random process. In a truly random process, no relationship exists between its current state and any future state. So when you observe that process and collect data on its current state, you can gain no understanding of that process that will allow you to predict anything about its future state. So no predictions can be made.

Our world is full of random processes intimately tied to nonrandom processes. Observations of those processes can be of high quality, but a random component adds an error to the data. That error makes the observations of poor quality, even though they have been made with high accuracy. The understanding that results will be unreliable not because of the observer, but because of the world itself.

Poor-quality predictions based on unreliable understanding are superstition, and superstition is a result of an inability to understand the world. Just as extremely high-quality predictions are great faith, extremely low-quality predictions are superstition or little faith. Those of little faith are unable to predict what will happen. So a failure to observe can be caused by the observer, and it can also be caused by the external world. The uncertainty brought on by this failure to observe is symbolized by darkness. The Bible writers personify that uncertainty as demons, devils, bad angels, or other similar things. Others personify that uncertainty as a characteristic of their gods: the gods are fickle and act at random to destroy men for no reason.

The Bible writers personify an "Ultimate Uncertainty" as Satan, who is the ultimate source of darkness in the world. Satan and his demons would mirror the UMkR and the angels, except that they de-energize objects in the world by extracting order from the world. Conversely, the UMkR would energize objects in the world by injecting order into the world.

For a living observer, the failure to observe leads directly to death. It is being in darkness. Since we all fail to observe to some extent, we are all in darkness to some extent. Being in darkness has three levels.

One is the sin of ignorance (not collecting data), which can lead to death. Another is the evil of closing your eyes to knowledge (deciding not to collect data), which can lead to death more quickly. The third level is the wickedness of embracing false knowledge (deciding to collect bad data), which can lead even more quickly to death. The prudent observer knows that there is error in every observation and seeks to characterize that error and to minimize it. From the perspective of the Bible writers, that process is a constant struggle against Satan and his demons. We know that Satan and his demons hate us because causing someone to fail to observe prevents them from having insights—it is the definition of hatred. Anyone who is the cause of sin hates those whom he causes to fail to observe.

With this framework, I think it's possible to understand the Bible and the riddles it contains much better than anyone has understood it in a very, very long time.

CHAPTER 4

The Story of the Flood

4.1 Understanding the Stories

GENESIS 5 IS A CALENDAR PUZZLE, but it is not a complete calendar. The remainder of the calendar is encoded in the story of Noah and his flood. This chapter will extract the remainder of the calendar. The story of the Flood begins in Genesis 6:1 and ends in Genesis 9:17. We're told that Noah built an ark to preserve his family and the air-breathing, land-dwelling creatures of the earth from a massive flood. On the day the Flood began, eight people entered the ark. The story used each person to represent a block of four years of the calendar, and a final 365-day year was added to create a 33-year calendar cycle. The last year of that 33-year cycle was the year of the Flood.

The story of Noah's ark and the Flood does tell us more about the calendar, but the calendar is a small part of the Flood story. This story contains an important lesson taught by our textbook Bible. We will cover the other important lessons these writers wanted to teach as well. The very first lesson in this ancient textbook describes the creation of everything, and the other lessons are about equally expansive subjects. The lesson that follows the creation teaches us how every kind of thing within God's creation was brought into existence. The next lesson tells us how living things came to be, and also how mankind was created. After that, a lesson explores the defining characteristics that separate mankind from the other animals. We are then instructed in a series of lessons that cover the history of man's "firsts" (such as the first sin or the first maker of iron tools). Each of these firsts is detailed to allow each student to *figure out for himself* how history unfolded from his distant past to his present.

Since these lessons were thought out so completely, they still have a lot to say to us about the subjects they cover. We still need to understand the nature of the universe in which we live, and we still need to understand what it means to be human. Adding the voices of these ancient writers to that conversation benefits us all. Human knowledge has increased greatly over the centuries, but basic human intelligence has not changed. So on these subjects, the authors could have a point of view that might benefit us as well. But deep within our philosophical hearts, we're sure that these "primitive" writers have little to offer modern thinkers like us. Our failure to see a calendar that is in plain sight tells us we have greatly underestimated the Bible writers. That underestimation has caused us to misunderstand their writings and has prevented us from understanding the region's history and our own. As we examine each of the stories from Genesis 1 to 11, we will learn that these writers were not the "primitives" we thought they were. We could, in fact, become the newest students of these ancient teachers.

4.2 Odd Contexts

The story of Noah's flood and his ark is complex. It is intertwined with the genealogy of Genesis 5, and that increases the difficulty of understanding both stories. I have separated out the genealogy of Seth and extracted the perpetual solar calendar it contained. Now we must examine the remainder of the story to see what is hidden within it. It is important to notice that during the Flood the earth was covered with water, just as it was in Genesis 1. Because of the clouds, the earth must also have been covered with darkness, as in Genesis 1. During the Flood, when the clouds parted and the light shone through, a "let there be light" moment occurred, just as in the Creation story. These correspondences show that the story of the Flood is, to some extent, a retelling of the story in Genesis 1. Yet the story of the Flood stands on its own.

The retelling of the Creation story of Genesis 1 within the story of the Flood is a little odd. It catches your attention. A calendar hidden within the genealogy of Seth is also very odd. The numbers catch your attention, but that's all they do until you figure out what they mean. Some of the other odd events in this story are all the talk about violence, giants, sex with angels, the Flood itself, and rainbows. From what we've seen above, these aren't actually mythological events but are put into the story to give us the clues we need to develop the insight to understand what this lesson is trying to teach us. Just as in the genealogy of Seth, we must first determine the subject matter of this riddle before we can ever make any sense of it. The list of odd things in this story is long, but the oddest thing of all occurs near the end, when Noah and the animals he's saved from the Flood come out of the ark: *Noah and his progeny are given the right to eat meat.*

When we consider this story and its context, the right to eat meat comes as a total and complete surprise. As far as we know, up to this point in the stories that make up Genesis 1 to 11, God has completely forbidden the eating of meat. If we look at the context of the story of the Flood, we see that violence was increasing in the world. We know that the first violent act was the murder of Abel by his older brother, Cain. Another part of the context was the relationship between God and man that wound its way from the Garden of Eden, to the translation of Enoch into heaven, to God promising never to destroy all life via another flood. Throughout this section of Genesis, the Bible writers tell us about all the "firsts" that occur. The first man, the first woman, the first sin, the first law, and the first mighty man (in the old style). Also in this list of firsts is the invention of musical instruments, cities, and bronze and iron implements. None of this seems to provide the slightest hint that God might grant man the right to eat meat.

After the Flood, it becomes clear that the culmination of the story is God granting to Noah and his sons the right to eat meat. After that grant, the writers tell an *extremely* odd story about Noah and his sons and one of his grandsons: Noah grew grapes and produced enough wine to become drunk. Noah's son Ham then shamed himself and his father when he walked in on a drunk and naked Noah. Later, in his anger, Noah issued a damning curse, but he did not curse his son Ham. He cursed Ham's son Canaan, Noah's grandson. Canaan was cursed to become the very first slave, and on this note this very odd story ends, and the more historical portion of the story begins.

4.3 The Right to be The Flood

Before the Flood, the text of the Bible says that the Flood would destroy all air-breathing, land-dwelling animals. God expands on this statement after the Flood when he says that the Flood destroyed "every living creature."[bb] Immediately after that, God gave mankind "everything" for food.[bc] It is certainly clear from the story that the Flood did not kill all fish, so it certainly did *not* kill "every living creature." Still, the authors tell us that the Flood killed everything, and then the authors say that God gave us everything for food. The two parallel statements are made in quick succession in the text. Those two statements are saying the same thing: The writers expressly state that the Flood killed every living creature, then they immediately say that mankind received the right to consume every living creature as food. This is the big clue.

When God gave Noah and his descendants "everything" for food, God had, in fact, granted Noah and his progeny the *right to be the Flood*. He commanded them to multiply and fill the earth, just as he had commanded the waters of the Flood. And it is hard to miss that the command for mankind to multiply and fill the earth was given at the very same time God gave Noah and his sons the right to eat meat. The Flood consumed all flesh on earth, and after God's command, humans had the right to multiply without limit and consume all flesh on earth. The authors clearly intended to link the two concepts with these parallel statements.

The insight that should immediately flash into our mind's eye is a single concept that merges the Flood and a large number of people. I merged the two ideas into a single concept of a "sea of people" all acting as one. So I concluded that *the waters of Noah's flood were people*.

Once we know that the waters of the Flood were people, the story can begin to make sense. At the beginning of the story the authors say, "When men began to increase on earth..."[bd] And at the end of the story the authors have God say, "Be fertile, then, and increase; abound on the earth and increase on it."[be] So what really happens in this story is that the "Flood People" fill the earth and consume every living thing they can find. After they have eaten everything, they die off. The die-off is told as the Flood subsiding. Then God gives Noah and Noah's progeny—the "New Flood People"—the right to do the same thing. So at the end of the story, the possibility of a new flood remains. To prevent the new flood from ever happening, God enters into a contract with his "New Flood People," and he signs it with a rainbow. The contract is designed to prevent the "New Flood People" from increasing without bound, consuming everything in sight, and dying off as the "Flood People" in this story did. So even though we do not yet know why God gave Noah and his progeny the right to eat meat, we do know that the story of Noah's ark is about people consuming the flesh of animals.

Dietary law is central to this story and to the overall story starting in Genesis 1. In the beginning, God created a strict hierarchy between plants and animals. All animals ate only plants. In the Garden of Eden, God cautioned Adam not to eat from the tree of knowledge of good and evil. When Adam ate from the tree anyway, God made him work very hard for his food, which still had to be "the plants of the field."[bf] When Noah came out of the ark and God gave him the right to eat meat or anything else, the world had made a quantum leap from being vegetarian to being omnivorous. If the waters of Noah's flood were people, then the Flood story has to be about lots of people eating lots of animals.

The change from vegetarianism to omnivorism is such a huge one that it must be covered in any lesson that is supposed to cover the history of mankind. Since this story is about people eating flesh for the very first time, it should explain how such a thing could happen, and why. So the story of the Flood is constructed to explain how the entire world moved from vegetarianism to omnivorism. The story also warns us that the consumption of flesh can have some very serious consequences.

The transformation from vegetarianism began in Genesis 6, where the authors introduce us to the Nephilim.[bg] The Nephilim were giants that resulted from sex between those known as the "sons of God" and the "daughters of men." It's not immediately clear what the writers meant when they used either phrase. We know that some people have decided that the untranslated "sons of God" phrase refers to angels, since it is construed to mean angels when translated elsewhere in the Bible. Others believe that the untranslated phrase refers to the progeny of Seth.[bh, bi] I think the phrase refers to angels, which leaves me with the awkward task of explaining what it means for angels to have sex with women. I will also have to explain why their offspring were giants. We determined what angels were in Chapter 3 and we will use that definition here.

What's actually happening in this part of the story is that the daughters of men are repeating the sin of eating a forbidden food. Eve's infamous sin was to eat what God had forbidden her to eat. That food was the fruit of a particular tree that grew in the Garden of Eden. In this case, the daughters of men were being seduced by the sons of God to eat a food that God had also forbidden, and that food was flesh. When the daughters of men ate flesh, it greatly improved their nutrition and made them much healthier. The women looked and felt much better. They continued eating meat because their better health made them more attractive. Their personal beauty is not a frivolous detail; it is an important part of the story. The daughters of men also fed the flesh that they ate to their children. The intense nutrition caused those children to grow much bigger, taller, stronger, and more attractive than others who were barely surviving on the meager nutrition they obtained from an earth God had cursed. (God cursed the earth to punish Adam and Eve.) The much taller and bigger children of these flesh-eating women became the Nephilim, and they were only "giants" when compared to their nutritionally deprived parents. They weren't giants, they were just big people.

The non-flesh-eating people like Noah and his children were threatened by what the Nephilim represented. As more and more of the children of the world were fed flesh, more and more people would become Nephilim as the older people died. The writers show that death was important to this process when they have God say that his spirit would not always blow inside man and thereby give him life, but that a man's lifetime would be 120 years long. In other words, those who followed God's law would die out and leave only the Nephilim behind. God's statement also implies that the consumption of flesh would result in people living much shorter lifetimes, since at this time people were living to be nearly a thousand years old. After the Flood, the recorded lifetimes rapidly decline to around 120 years. Everyone on earth would one day be Naphil. The only person refusing to eat meat was Noah. He and his family were the only vegetarians left.

I want to emphasize that the story of Eve being seduced by the serpent in the Garden of Eden and the daughters of men being seduced by the idea of eating flesh are the same story. In the Garden of Eden story, the forbidden food was the fruit of the tree of knowledge of good and evil, and in this story the forbidden food is flesh. In this story, we have the daughters of men; in the Garden of Eden story, Eve is the daughter of Adam—she is the daughter of a man.

The authors tell us that the Nephilim were the "mighty men of old."[bj] But after the Flood was over, they tell us that Nimrod was the "first mighty man." When you read about Nimrod, it becomes clear that he was a mighty man because he was a great hunter. As such, he presumably ate the flesh of the animals he killed (they fell before him). So when the authors tell us that the Nephilim were mighty men, the direct implication is that they were also hunters. Of course, they would need to be hunters only if they ate the flesh of animals. So when the writers refer to the Nephilim as mighty men, they confirm that this story is about eating flesh. In fact, one suggested meaning of the word Naphil is "to fall,"[bk] as in "to fall upon," which is frequently interpreted as referring to war, but can also refer to hunting, since a hunter falls upon his prey, and it falls before him. So a Naphil is a hunter and not necessarily a giant. The hunting would have come first, so the first Nephilim would have been the same size as everyone else. Their children, who also hunted, would have become much larger, and their size would have contaminated the original meaning of the word.

The idea that the Nephilim were somehow nonhuman is incorrect. They were human beings just like everyone else. They were just big. They did, however, achieve that size by violating God's dietary law of vegetarianism. So their size was direct evidence that they were breaking God's law. It was evidence of their wickedness. The fact that they otherwise would have had great difficulty feeding themselves did not excuse their sin. After the Flood, God allowed everyone to eat meat, so Noah's grandchildren became "giants" as well. These were the mighty men of Nimrod's generation. They were the "New Nephilim," the "New Hunters," or the "New Flood People," and so are we.

Because of this story, exceptional stature became associated with violating God's dietary laws. Long after the Flood, anyone who achieved exceptional stature was therefore presumed to have achieved it by some sort of wickedness and would deserve to be destroyed for it. Because Noah's descendants were already of our stature, when they encountered giants they were, in fact, encountering giants among the giants.

Killing to eat fills the earth with violence. When the Nephilim (hunters) and the daughters of men thought about their next meal, from God's perspective they were having evil thoughts. Later on in the narrative, the writers reveal that all flesh had corrupted itself, not just mankind. Since the corruption of mankind at this point involved eating flesh, we have to conclude that the other animals had corrupted themselves by eating one another.[bl] This turn of events was too much for God to bear, and he determined that he would destroy the earth through this "Flood of Nephilim."

This may also be where the concept of a "clean" animal originates. I've always thought that the reason an animal was called "clean" or "unclean" in the Bible was to indicate the likelihood that someone would contract a disease when the flesh of the animal was consumed. At the point where the word is first used to classify animals, this cannot be what the phrase means. Since the other animals had just begun to violate God's law concerning the consumption of flesh, it would make sense that a clean animal would be one that remained a vegetarian and did not violate God's law. An unclean animal would be one that had corrupted itself by eating flesh in some way. Later, when humans were allowed to eat animals, whether the animal remained vegetarian and did not violate God's law might still have played a part in its classification as clean or unclean.

As far as we know, the idea of clean and unclean animals did not exist when Abel sacrificed an animal to God (though it appears that a religious culture was already in place). At that time, all animals still ate plants, and the story implies that humans did so as well. By the time of Noah, humans ate animals, and animals ate one another. When Noah entered the ark, "clean" could mean that an animal remained vegetarian, as Noah and his family did. After the Flood, "clean" could mean that the animal remained vegetarian, and it could also mean that the animal was suitable for human consumption. When the authors use the phrase "clean" to describe the animals, they are hinting again that the story of the Flood is about the consumption of flesh.

What this story is actually describing is a view of how humans moved from an idealized world of vegetarianism to the world as it was in the authors' time, and as it is today. They tell the story via an idealized version of a population explosion and crash that certainly can and has happened in human populations. In their version, an original human population moved from a land of plenty (the Garden of Eden) to a region that was much less accommodating. Over time, this group of people suffered from poor nutrition, which ultimately caused their offspring to suffer from small stature.[bm] At some point, the people began killing and eating animals. Their new lifestyle in their new environment was unsustainable, but they were not aware of that. The increase in nutrition led to an increase in fertility and fecundity. Improved nutrition also allowed children to exceed the stature of even their well-fed vegetarian ancestors. The resulting population explosion and famine left only a few people alive at its end. Noah saw the coming population explosion and famine and took measures to protect himself, his family, and his resources from this Flood of humanity. And so, we have the story of the ark.

4.4 The Ark

Many people continue to question why the word "ark" is used in the Bible to describe Noah's ship. Two other important objects in the

Bible are described using the word. When Moses was an infant under threat of death, his mother placed him in an ark and floated it in the Nile River in an attempt to save his life. When Moses led the Israelites out of Egypt, he had an ark built to hold the tablets of the law written by the hand of God. The preciousness and fragility of the contents are characteristics shared by each ark. Precious and fragile items must be carried in a strong container designed to protect what's inside. That container is the ark.

Another common feature of each ark is its removable cover. If we used these common features to identify something we would recognize today, we would be describing a shipping container. For us, a shipping container is a rugged enclosure designed to take significant abuse. It's designed to protect its contents and to be opened and closed easily.

It's also possible to add protection to an ark to enhance its protective abilities. Moses' ark was coated with pitch to protect the baby Moses from the water of the Nile River. The Ark of the Covenant was overlaid with gold on the inside and outside to greatly increase the lifetime of the wood from which it was made. That same gold overlay could also have helped protect the contents by helping to seal out a hostile environment.

When God told Noah how to build his ark, God commanded him to coat it with bitumen on the inside and on the outside. Bitumen is a very thick petroleum product that seeped from the ground in the geographical region of lower Mesopotamia. Since this story takes place somewhere in that region, bitumen would have been readily available. Lower Mesopotamia consists of modern-day southern Iraq and portions of Kuwait and southwestern Iran near the city of Ahvaz, located near the ancient plains of Susa.[bn] In Noah's time, boat builders and maintainers used bitumen to seal boats and keep them watertight. So when the authors tell us that God instructed Noah to seal the ark with bitumen, that instruction is consistent with the story and its location.

Today, many of us use the word "asphalt" to refer to bitumen. They are essentially the same thing. The ancient Mesopotamians obtained bitumen from seeps, and today asphalt is produced as the residue of crude-oil distillation. The ancient Mesopotamians used bitumen extensively. It was famously used as mortar in the construction of important buildings and temples.

Noah was commanded to build an ark so that he and his family would survive the coming deluge. That deluge was a population explosion that would lead to the very first war. Noah was barricaded in the ark, and the Nephilim were outside. Noah's war, as the first war, was not a war in the way that we understand it. Since there was no actual water in the Flood, Noah's ark was not an actual boat. It was a *fortified longhouse*[bo] built as a *hill fort*.[bp, 23]

A longhouse was an ancient dwelling that is reported to have originated within farming communities about 4500–3000 BCE. Family groups and their animals all occupied a single, multistory dwelling, or longhouse.[bq, 24] Some European forms of the longhouse were long, windowless buildings covered with a roof in which there was an opening for smoke to escape. In addition, a longhouse could have a door in its long side. The longhouse was a place for both humans and animals to live, and in some versions it was divided into three compartments.[br] A hill fort was an ancient fortification strategy that placed a fortification on a low hill. All of these features are characteristics of Noah's ark (except that the three "storeys" of the ark might correspond to actual levels in a multistory longhouse or the three compartments into which some types of longhouse were divided).

So Noah built the first *fort,* and the story of his voyage is the story of the first *siege*.[bs] With this story, the Bible writers are exploiting an analogy between a fort under siege and a boat on the ocean besieged by waves. The very first fort wouldn't be much of a fort. Of necessity, it would be a fortified version of a building normally used for other purposes. A communal house, such as a longhouse intended for housing people, storing food, and housing animals, would be a natural place to seek refuge. Such a building would have been designed to protect its valuable contents and occupants from severe weather or from raids by neighbors. It would be an ark. The fortification of the building did not have to be perfect, and we would expect that the first fortification would not have been very sophisticated. The ultimate goal of any attacker would be to capture some percentage of the resources stored in the fortified storage building. Any attack that resulted in the destruction of all the resources in the fort would be counterproductive. So using fire, for

---

23 This fort would be of wood, however.
24 Large buildings with multiple rooms that housed multiple families existed in Jafarabad, just east of Mesopotamia, in ancient times.

instance, to destroy the fort would be something that rational attackers would avoid. The occupants of the fort would have to deal with the fact that people are frequently not rational, however.

We know that materials were available to make bitumen relatively fire-resistant. It was common in ancient Mesopotamia to mix bitumen with other materials. Bitumen that has been recovered from that time is mixed with sand and other fillers.[bt] We also know that bitumen was mixed with powdered limestone.[bu] Mixtures of bitumen, called mastics, might be more resistant to fire simply because one of the components is a highly nonflammable material. For instance, when we mix asphalt (bitumen) with mineral aggregates to pave our roads, the combination is much less flammable than bitumen alone.[bv] In addition, powdered limestone (containing calcium carbonate), when heated sufficiently, will decompose and emit carbon dioxide. The carbon dioxide emissions would tend to suppress the fire that caused them. Today we know that a mixture of bitumen and powdered sodium bicarbonate (baking soda) has been shown to resist fire in exactly this way.[bw] When ignited, the burning mixture easily emits carbon dioxide and forms a residue (called a "char") over the burning bitumen. Together, the carbon dioxide and the char act to help extinguish the fire at its source. In ancient Mesopotamia the mineral trona was available and could have been used instead of sodium bicarbonate. When trona is heated to decomposition, it releases carbon dioxide in a manner very similar to sodium bicarbonate.[bx] Powdered trona mixed with bitumen should resist fire much more effectively than bitumen alone or bitumen mixed with other more stable minerals.

We also know that the mineral natron was available in Egypt. Natron is similar to both sodium bicarbonate (as nahcolite) and trona (the three mineral salts are deposited in similar ways).[by] The ancient Egyptians famously used powdered natron as a desiccant in the process of mummification. A mixture of powdered natron and bitumen should form a mastic with properties similar to those containing either sodium bicarbonate or trona. When the heated mixture is applied to a building and allowed to harden, it should form an excellent protective coating. So when Noah coated the ark on the inside and outside with bitumen, it's easy to imagine that he was fireproofing, waterproofing, and armoring the building.

The authors tell us that Noah built the ark 30 cubits high, 50 cubits wide, and 300 cubits long. The text of the Bible says that the waters covered the hills 15 cubits deep. Since the ark is a building with walls and a roof, climbing the walls would be the primary way to breach the fort's defenses. So the characteristic of the Nephilim that would have been the most important would be their height from ground to upstretched arms. That height is what a single person could easily climb. Now, it makes no sense that Noah would know how high above the hilltops the water would be if the water was real and there was a real flood. It also makes no sense that hills of differing heights would be covered by water to the same depth, unless the water was clinging to the hills like a "sea of people" would cling to the hills. And it is also clear that if the enemy can easily scale a 15-cubit wall, making the ark 30 cubits high would be a good defensive strategy for a land-based fort. But it's a very bad situation if you've built a real ship that is 30 cubits deep and the mountains are covered by only 15 cubits of real water. In this way, the authors confirm to us that the ark is not a real ship, and the water is not real water.

But the other thing that the puzzle authors do by giving us the dimensions of the ark is to let us know that the ark is 700 cubits around its perimeter. When added to the ark's 30-cubit height, we get 730 cubits. Now, this number is not that familiar to us, but we have already established that the puzzle authors double numbers to obscure their significance. So we note that $730 = 365 \times 2$, and this fact much more directly implies that the ark is symbolic of a 365-day year. As we noted earlier, our calendar is composed of 4-year periods consisting of three 364-day Divisible Years, followed by a 369-day Leap Year. The puzzle authors indicate that we should use eight of these 4-year periods in our calendar when eight people enter the ark before the Flood. It was my contention that we know that the 33rd year is 365 days long, because Noah was described by the phrase "Noah walked with God," and this phrase is also used to describe Enoch, who lived exactly 365 years. When the authors make the ark symbolic of the 365-day year, they confirm to us that the last year in the 33-year cycle is 365 days long.

Since a cubit is defined as the length of a person's arm from the elbow to the tip of the middle finger, each person has his own cubit. In a particular population, an average or accepted cubit can exist as long as the people are not of radically differing sizes. Since the Nephilim were much larger than their parents, who were Noah's size, the cubit for the Nephilim should be much larger as well. Since *we* are the New Nephilim, we should be about 15 cubits tall with our arms reaching above our heads. A good average reach for a tall modern man is 8 feet (2.4 m) with arms stretched above the head. If we divide 8 feet by 15 we find that Noah's cubit would be about 6.4 inches (163 mm) in length. On average this is around 1/3 the size of most ancient cubits. An actual ark based on this cubit would be 160 feet (48.77 m) long, 26.7 feet (8.13 m) wide, and 16 feet (4.88 m) high. These measurements are in line with the largest ancient European longhouses.[bz]

We don't know how tall Noah would have been, based on this story, but as a practical matter, if the Nephilim were much better nourished than their parents, they might be 50% to 100% taller than their parents. So on average their parents could be 7.5 to 10 cubits tall with arms reaching above their heads (of course, I presume that basic body ratios did not change with size). For us, this would be 4 feet (1.22m) to 6 feet (1.83m). Noah and others of that time could have been really quite small. According to the story, the small stature would have been caused by the curse God placed on the Earth after Adam and Eve violated God's law in the Garden of Eden. So Adam and Eve would have been taller—maybe not as tall as the Nephilim, but not as small as Noah.

The size of Noah's cubit, the size of Noah's ark, and the estimates of Noah's height are necessarily very rough approximations, but a structure based on these estimates could certainly have been built on a long mound or low hill. That ridge would be Mount Ararat, where the ark supposedly settled during the Flood. Ararat means "high land,"[ca] and the authors seem to mean only that. And we know that some societies in the area at that time built large houses or major buildings on platforms at the center of their settlements.[cb]

The clear similarity between building a structure on "high land" in the center of a settlement and the story of Noah's ark suggests that this story may be trying to explain the origin of that practice. The earliest buildings may have been built on naturally occurring rises, and later buildings may have been built on man-made mounds. Ultimately the mounds could have evolved into higher and more elaborate constructions until ziggurats such as the Tower of Babel arose. So the story of the Tower of Babel would naturally follow and extend the story of Noah's ark. I think it is clear that Noah's ark/fort need not have actually existed for this story to be told.

To survive the siege, the people in the fort must have water and food for the duration. We don't have any information on how Noah provided water, and I suspect that particular detail, if provided, would give away the secret of the ark. It would be easy to dig a well if the ark was actually a fortified building. To get food for the siege, God told Noah to store up "every type of food that is eaten,"[cc] which is a wondrously pregnant statement. In a purely vegetarian world, all food would be plant material. But in an omnivorous world, some animals would be food for others. The authors have God make the "every type of food" statement to clearly imply omnivorism without giving away the secret of the story.

Once the population exploded and all available resources in the region were consumed by this Flood of humanity, Noah's fortified building would have to resist all attempts to break in. As a known source of stored food, Noah's ark would be a prime target for any raid. When food resources were completely exhausted, the hungry masses would assemble outside Noah's ark and attempt to break into it. The "waters of the Flood" would "besiege" the ark like a ship on the ocean. As time passed, the people assembled outside the ark would begin to starve and die. The die-off would represent the "receding" of the "waters of the Flood." Noah could have shared some of his food with some of the people outside the ark, but he certainly could not have fed them all. Ultimately, Noah would have had to attack the more organized groups within the assembled masses to preserve the integrity of the ark. From Noah's perspective, however, the people assembled outside the ark brought this calamity upon themselves by breaking God's law. According to the story, the siege of Noah's ark lasted for a little more than a year. And as you might expect, keeping time is very important during a siege.

4.5 The Flood and the Calendar

When the authors of the Flood story give us precise dates for the major events of the Flood, they are clearly trying to tell us something. Noah entered the ark on the 10th day of the Second Month, and seven days later, on the 17th day of the Second Month, the Flood began. The calendar we extracted from the genealogy of Seth contained no months. While the year could easily have been divided into 13 28-day months, the Bible writers did not give any indication that they actually subdivided the year in that way. We've noted that the story of Noah's ark and the genealogy of Seth are intertwined. The authors interwove these stories because they are closely related. The dates given in the Flood story allow us to extract the intra-year divisions of that calendar. The Flood story also tells us where we should put the intercalary days[25] associated with the three different years used in this calendar. It further provides a rationale for locating the epagomenal days[26] associated with a month that has a length that doesn't evenly divide into the year.

Before we take on the task of extracting more of the calendar from this story, let's review the three different years used in this calendar. The Perfect Year is 365 days in length. A year of this length occurs once every 33 years. The Divisible Year is 364 days in length and is the common year. The Leap Year is 369 days in length and occurs once every four years. It is the perfect 365-day year that is detailed in Noah's flood. We can remove an intercalary day from the 365-day year to get the 364-day year. We can add five intercalary days to the 364-day year to get the 369-day year. Using the 365-day year, I have diagrammed every major event of Noah's flood in Figure 4.1.

---

25 Intercalary days are days added to the common year to create years of different lengths. The years of different lengths are added to the calendar to synchronize the year with the seasons or some other astronomical event. For us and our Gregorian calendar, when we add February 29th to the calendar every four years, that day is an intercalary day, and it produces a 366-day year. The word "calary" implies calendar.

26 Epagomenal days were days added to the Egyptian calendar that were not a part of any month. They also served to synchronize their 360-day calendar to the solar year. For our use, an epagomenal day is one of a group of days added to the calendar to synchronize it to an astronomical event like the solar year. Epagomenal days are different from intercalary days, because intercalary days create years of different lengths. Epagomenal days, as defined here, would complete the common year.

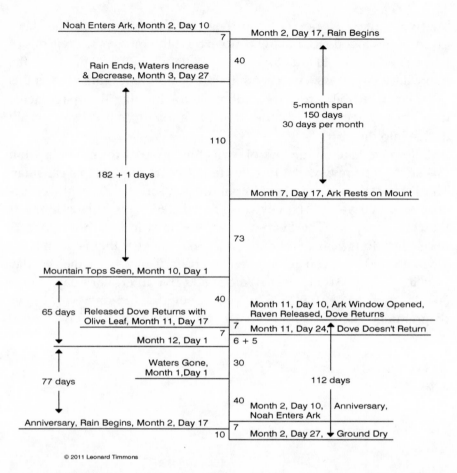

© 2011 Leonard Timmons

Figure 4.1: Flood Events – Perfect Year. The major events of Noah's flood arranged from beginning to end. Some dates are specified in the text of the Bible, other dates are calculated. The year is 365 days long.

The time line begins when Noah enters the ark. He entered the ark with seven other people on the 10th day of the Second Month. After seven days the rain began, and the text specifically marks that day as the 17th day of the Second Month. Then there is a 40-day period of rain on the earth. Next there is a 150-day period where the waters rise and then subside. It's easy to read this as a 40-day period followed by a 150-day period. But that's not what happened. The 40-day period of rain is included *within* the 150-day period because the rain is what makes the waters rise. So there is a 40-day period of rain and increasing waters, and a 110-day period where the waters stabilize and then decrease. The ark rests on Mount Ararat on the 17th day of the Seventh Month. From the 17th day of the Second Month to the 17th day of the Seventh Month is exactly five months, exactly 150 days. This is clearly the puzzle authors' way of telling us that they have divided the year into months that are 30 days long.

Once we know that each month is 30 days long, we can calculate the date on which the 40 days of rain ended. That date is the 27th day of the Third Month. The next major event occurs when Noah sees the mountain tops on the first day of the Tenth Month. Since we know that the months are 30 days long, we can calculate that the mountaintops were seen 73 days after the ark rested on Mount Ararat. The next major event in the time line occurs 40 days later when Noah opens the window in the roof of the ark and releases a raven and a dove. The dove returned, but the raven did not. The date of this event is the 10th day of the Eleventh Month. Seven days after that, on the 17th day of the Eleventh Month, Noah released a dove, and it returned with an olive leaf in its beak. After another seven days, on the 24th day of the Eleventh Month, Noah released a dove, and it did not come back.

At this point in the time line, six days remain in the Eleventh Month. We also know that 12 months of 30 days adds up to 360 days, which is five days short of a 365-day year. We can make the year 365 days long by creating a block of five days that will not be in any month (epagomenal days). We must decide where to put those days in the calendar. Since months are based on the cycles of the moon and the moon goes through a cycle that is approximately 29.5 days long, 12 months of 29.5 days is 354 days. A lunar year based on this kind of calculation is approximately 11 days shorter than the solar year. If we put the 6 days left in the Eleventh Month together with the five epagomenal days that will not be a part of any month, we will collect 11 days together at the end of the Eleventh Month. Doing this allows us to capture the lunar year beginning at the first day of the Twelfth Month to the 24th day of the Eleventh Month with 11 extra days to synchronize the lunar year with the solar year. This design causes *the first day of the Twelfth Month to be the start of the year.*

The drying of the waters off the earth is the next major event that occurs in the time line. That event occurs on the first day of the First Month. For the reasons mentioned above, this day is not the beginning of the year, though it is the beginning of the calendar. The first day of the First Month is 41 days from the 24th day of the

Eleventh Month (the day Noah released the dove that did not return). After 40 more days, on the 10th day of the Second Month, the anniversary of Noah's entering the ark occurs. Seven days after that, on the 17th day of the Second Month, the anniversary of the beginning of the Flood occurs. Neither of these two events is explicitly marked in the text. I have included the anniversaries so I can show that those dates are related to the beginning of the year on the first day of the Twelfth Month. The final event in this sequence occurs when the ground becomes completely dry on the 27th day of the Second Month, and the Flood is at an end.

The authors did not specify the number of days between the resting of the ark on Mount Ararat and the first day of the Tenth Month. So I calculated that there were 73 days between the 17th day of the Seventh Month (when the ark rested) and the first day of the Tenth Month. I had already determined that there were 110 days between the 27th day of the Third Month (at the end of the 40 days of rain) and the 17th day of the Seventh Month. The length of this entire timespan is 183 days, or just about one half year. See Figure 4.1.

You might remember that the number 182 was central to the genealogical calendar. It's important here as well. For reasons that will become apparent, I show the 183-day span as 182 + 1 days. In addition, a 112-day span starts at the 10th day of the Eleventh Month (when Noah opened the window[27] in the roof of the ark) to the 27th day of the Second Month (when the ground became dry and the Flood ended). The number 112 is another reference to the genealogical calendar, where 112 = 2×56 and the number 56 and its reverse, 65, are so very important. A 77-day span begins on the first day of the Twelfth Month (the beginning of the year) and ends on the 17th day of the Second Month (the anniversary of the day the rain began). With this information, it becomes clear that Noah entered the ark on the 70th day of the year, and the rain began on the 77th day of the year.

The year of Noah's flood is 365 days long. One of the great things about the Flood story is that it allows us to adjust the year for the intercalary days required by the changing length of the year. For the 364-day year, we simply make the 183-day interval 182 days long. When intercalary days must be added or removed, we will adjust the length of the 73-day interval. The number of days in this period was

---

27 Probably a covered chimney.

purposely left undefined within the story and that strongly implies that intercalary days should be added or removed from this interval. We could add or remove days at the end of the Seventh, Eighth, or Ninth Month. I chose to add and remove days at the end of the Ninth Month because the most logical place to add or remove days is at the end of the 73-day period.

Other ancient calendars in this region intercalated in the seventh month.[cd] For those calendars, the beginning of the year occurred at the spring equinox, so their seventh month would roughly correspond to the Ninth Month of this calendar. Since we're modifying the Ninth Month here, we will remove one day to give it 29 days and produce the Divisible Year. That year is the common year for this calendar. See Figure 4.2.

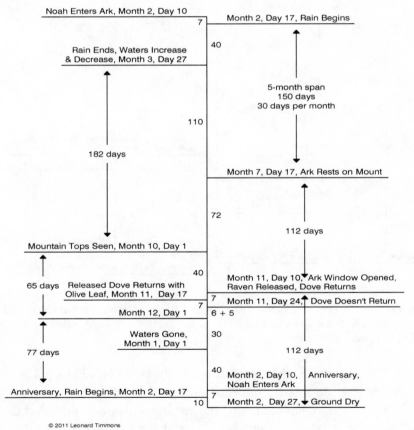

© 2011 Leonard Timmons

Figure 4.2: Flood Events – Divisible Year. The major events of Noah's flood arranged from beginning to end. Some dates are specified in the text of the Bible, other dates are calculated. The year is 364 days long.

The 369-day Leap Year is produced by adding four intercalary days to the Perfect Year, or five days to the Divisible Year. It also makes sense to place those days between the Ninth Month and the Tenth Month to create a second 77-day period. See Figure 4.3.

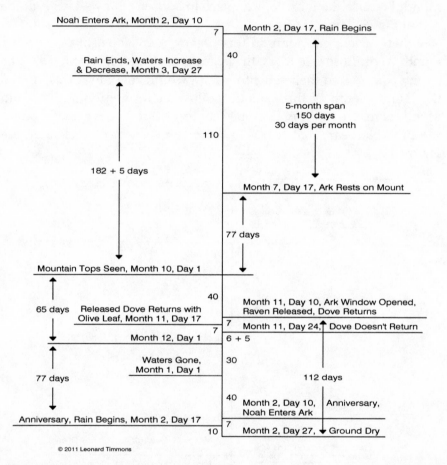

© 2011 Leonard Timmons

Figure 4.3: Flood Events – Leap Year. The major events of Noah's flood arranged from beginning to end. Some dates are specified in the text of the Bible, other dates are calculated. The year is 369 days long.

The 77-day periods that flank the 65-day period in Figure 4.3 are an echo of the same mechanism used in the original genealogy puzzle. This mechanism confirms that we have properly decoded this calendar and placed the intercalary days at their proper places. Note also that our original 183-day span is flanked by two 40-day intervals. That span changes to 182 days for the 364-day Divisible Year and to 187 days for the 369-day Leap Year. Note also that the number 187 appears prominently in the genealogy of Seth, along with 182, 65, 112, and 77. So these are the natural divisions of the calendar into numbers that have significance.

As a final indication that we have the correct solution to this puzzle, we note that the Flood does not last exactly one year. This story lasts exactly 382 days when the base year is 365 days long. Of course 382 is 182 + 200, with the implication that it is easy to divide the 200 into two 100-day intervals, thus repeating the theme of a time period flanked by two equal time periods.

The number 11, the difference between the length of the lunar year and the solar year, occurs in a couple of places in interesting ways. First, consider the 6 + 5 = 11 days at the end of the Eleventh Month. Technically I put those days there, but I think the structure of the puzzle insists that they be inserted at that location. The 6 and the 5 are 65 and 56 and were extremely important in the first puzzle. The other obvious 11 occurs in the 110-day period when the ark was afloat.

With this information, we can produce an example calendar for a 365-day Perfect Year, show the loss of a day for the 364-day Divisible Year, show the intercalary days for the 369-day Leap Year, show the epagomenal days for a year with 30-day months, and show the lunar year overlay. See Table 4.1.

Table 4.1: Divisible Calendar

This calendar has monthly epagomenal days that don't appear in any month. It also contains a lunar year that begins with the Twelfth Month and ends on the 24th day of the Eleventh Month. The three year-lengths of 365 days in the Perfect Year, 364 days in the Divisible Year, and 369 days in the Leap Year are represented as well. The intercalary days associated with the changing year length are added at the end of the Ninth Month and between the Ninth and Tenth months.

MONTH TWELVE

Sunday	Monday	Tuesday	Wednesday	Thursday	Friday	Saturday
1 YEAR BEGINS	2	3	4	5	6	7
8	9	10	11	12	13	14
15	16	17	18	19	20	21
22	23	24	25	26	27	28
29	30					

MONTH ONE

Sunday	Monday	Tuesday	Wednesday	Thursday	Friday	Saturday
		1 WATERS GONE	2	3	4	5
6	7	8	9	10	11	12
13	14	15	16	17	18	19
20	21	22	23	24	25	26
27	28	29	30			

Table 4.1: Divisible Calendar Continued

MONTH TWO

Sunday	Monday	Tuesday	Wednesday	Thursday	Friday	Saturday
				1	2	3 Enters Ark
4	5	6	7	8	9	10 Rain Begins
11	12	13	14	15	16	17
18	19	20	21	22	23	24
25	26	27 GROUND DRY	28	29	30	

MONTH THREE

Sunday	Monday	Tuesday	Wednesday	Thursday	Friday	Saturday
						1
2	3	4	5	6	7	8
9	10	11	12	13	14	15
16	17	18	19	20	21	22
23	24	25	26	27 Rain Ends	28	29
30						

MONTH FOUR

Sunday	Monday	Tuesday	Wednesday	Thursday	Friday	Saturday
	1	2	3	4	5	6
7	8	9	10	11	12	13
14	15	16	17	18	19	20
21	22	23	24	25	26	27
28	29	30				

Table 4.1: Divisible Calendar Continued

MONTH FIVE

Sunday	Monday	Tuesday	Wednesday	Thursday	Friday	Saturday
			1	2	3	4
5	6	7	8	9	10	11
12	13	14	15	16	17	18
19	20	21	22	23	24	25
26	27	28	29	30		

MONTH SIX

Sunday	Monday	Tuesday	Wednesday	Thursday	Friday	Saturday
					1	2
3	4	5	6	7	8	9
10	11	12	13	14	15	16
17	18	19	20	21	22	23
24	25	26	27	28	29	30

MONTH SEVEN

Sunday	Monday	Tuesday	Wednesday	Thursday	Friday	Saturday
1	2	3	4	5	6	7
8	9	10	11	12	13	14
15	16	17 Ark Rests	18	19	20	21
22	23	24	25	26	27	28
29	30					

Table 4.1: Divisible Calendar Continued

MONTH EIGHT

Sunday	Monday	Tuesday	Wednesday	Thursday	Friday	Saturday
		1	2	3	4	5
6	7	8	9	10	11	12
13	14	15	16	17	18	19
20	21	22	23	24	25	26
27	28	29	30			

MONTH NINE
Divisible Year has 29 days, Perfect Year and Leap Year have 30 days.

Sunday	Monday	Tuesday	Wednesday	Thursday	Friday	Saturday
				1	2	3
4	5	6	7	8	9	10
11	12	13	14	15	16	17
18	19	20	21	22	23	24
25	26	27	28	29	**30**	

LEAP YEAR INTERCALARY DAYS[28]
In Leap Years, four additional days are not in any month.

Sunday	Monday	Tuesday	Wednesday	Thursday	Friday	Saturday
						1
2	**3**	**4**				

---

28 These days may be placed anywhere after the 17th of the Seventh Month and before the
    1st of the Tenth Month. The other logical places to remove and add days would be at the
    end of either the Seventh or Eighth months.

Table 4.1: Divisible Calendar Continued

MONTH TEN

Sunday	Monday	Tuesday	Wednesday	Thursday	Friday	Saturday
						1 Mountain Tops
2	3	4	5	6	7	8
9	10	11	12	13	14	15
16	17	18	19	20	21	22
23	24	25	26	27	28	29
30						

MONTH ELEVEN

Sunday	Monday	Tuesday	Wednesday	Thursday	Friday	Saturday
	1	2	3	4	5	6
7	8	9	10 Dove & Raven	11	12	13
14	15	16	17 Dove & Olive	18	19	20
21	22	23	24 Dove Gone	25	26	27
28	**29**	**30**				

MONTHLY EPAGOMENAL DAYS

Twelve months of 30 days is 360 days. Five epagomenal
days are added to complete the year.
The last six days of Month Eleven and these five days are the eleven days that synchronize
the lunar year with the solar year.

Sunday	Monday	Tuesday	Wednesday	Thursday	Friday	Saturday
			1	**2**	**3**	**4**
5						

Although this calendar is not one that we know was used in history, it seems clear that some ancient calendar designer created it or something very similar to it. This story itself mentions a calendar, but does not describe it at all. We don't know whether the calendar we've extracted is that calendar (though I think it likely is). Nor do we know whether generating this calendar was just an intellectual exercise intended to be discovered by those who were exercising and expanding their insight, or whether it was a calendar used only by the Bible writers, or both.

You might notice, however, that our month of January (using the Gregorian Calendar) roughly corresponds to Month Twelve. And our month of December roughly corresponds to Month Eleven. If there is a true correspondence, then the spring equinox would occur on the 77th day of the year, or Day 17 of Month Two, when the rain that starts the Flood begins. It is also hard to ignore the period beginning on the 25th day of Month Eleven, which would correspond to December 25 in our calendar. These are the days that synchronize the lunar year with the solar year, and they can be viewed as being dedicated entirely to the sun. We celebrate Christmas on the 25th of December, and other cultures have celebrated various festivals on this day throughout history. It could be possible that this calendar is one of the early reasons why this day and the 10 days that follow it might be considered special. We don't really know the age of this calendar, so it is also possible that this calendar inherits these characteristics from earlier calendars.

4.6 After the Flood

I mentioned above that at some point during the siege, Noah would have had to fight back against organized groups within the mass of people besieging his fort. The authors give us the details of that fight when they tell us about Noah releasing a series of birds. When Noah first opened the window in the roof of the ark, he released a raven, an ancient symbol of war. The release of the raven represents Noah's active fight against the Nephilim. We also know that the Nephilim (symbolized by the waters) were already decreasing from the earth, and we concluded that they were failing to breach the ark's defenses and dying of starvation. Since the raven does not return to Noah in the ark, his release of the bird represents a campaign against the Nephilim that began when the raven was released.

Noah released a dove at the very same time he released the raven. Since the dove is an ancient symbol of peace, we would expect that the authors are saying that Noah made overtures of peace to the Nephilim at the same time he started his campaign against them. The nearly immediate return of the dove tells us that Noah's peace overtures were rejected. Since the raven was still "out there," the campaign against the Nephilim was still going on. During this interval, Noah's campaign against the Nephilim might have killed significant numbers of them. Noah's second release of the dove represents a second overture of peace. When the dove returned, it had in its beak an olive leaf, yet another ancient symbol of peace. The leaf represents the willingness of the remaining Nephilim to accept Noah's terms of peace. Noah then released another dove that did not return, representing the Nephilim's acceptance of Noah's terms of peace.

The end of the Flood occurs when all the water has completely dried from the earth. The waters are the Nephilim, so they must disappear as the waters dry, and a dry earth means that the Nephilim are completely gone. Since this story is modeled on a population explosion and collapse, we would not expect everyone to die, even if there was a war associated with the population collapse. The loss of every Naphil could mean that they died to the very last person, but that is not what the writers are saying here.

To understand what the authors are trying to tell us, we must first understand their concept of succession. Succession is an important concept that is used in many places and is handled as a father-son relationship. The father is succeeded by the son. The father dies, the son lives. But the father and son are, to some extent, the same person. This concept can be used wherever one thing succeeds another. It can be used whenever a particular thing changes into a similar-but-different-thing. The original thing would be the father, and the new thing would be the child. The original thing dies, and the new thing lives just as father and son do. (This concept also encompasses the idea of death and resurrection. The father dies but lives again within the son.) When the Bible writers use this concept, they invoke the idea of sonship by using the term "son of." So we have man and the son of man, the prophets and the sons of the prophets, God and the sons of God, etc.

In this case, the sonship of succession is also present. When the Nephilim accepted Noah's terms of peace, they were transformed into something new—New Nephilim. The original Nephilim "died," and the New Nephilim (or the "sons of the Nephilim") began to live. Sonship differs in this case because we have two fathers and one son. The Nephilim are one father, and Noah is the other. The New

Nephilim became Noah's descendants because they accepted Noah's terms of peace. This act made them Noah's successors, and Noah became their "father." The story implies that Noah's descendants were not just his children, but everyone who accepted his terms of peace. All the Nephilim who did not starve or were not killed in the fighting became New Nephilim when they accepted Noah's terms of peace. The Nephilim were also the fathers of anyone who consumed flesh. So with the compromise, Noah and his descendants became the "sons of the Nephilim" or the New Nephilim. This compromise caused the "waters," the Nephilim, to disappear from the face of the earth; the earth was completely dry.

The peace treaty Noah made with the Nephilim is detailed in Genesis 9. This is where God speaks to Noah and grants him the right to eat meat. The dietary rules that were established are:

I)   You can eat anything, but:

II) You cannot eat an animal while it is alive.

III) You cannot eat human flesh, nor can any animal eat the flesh of a man.

The need to make these restrictions tells us about the level of depravity the Nephilim had reached before and during the Flood. Even so, the desire to eat flesh remains evil, and that desire continues to corrupt mankind. God decided to overlook mankind's evil acts if we would keep the dietary rules above. Should we agree to observe those rules, he promises to never again destroy all flesh on earth, even though destruction should be the penalty for eating flesh.

Noah also sacrificed a number of animals when he and his family came out of the ark. The sacrifice tells us that the consumption of flesh would ultimately be mediated by the rules governing sacrifice. This story and the story of Abel and Cain seem to imply a preexisting set of rules. The details of those rules, whether they were cultural norms or specific laws, are not mentioned in either story, however.

Noah was born to deliver mankind from the curse that God placed on the earth after the sin of Adam and Eve. When Noah was born, his father, Lamech, prophesied, "This one will provide us relief from our work and from the toil of our hands, out of the very soil which the LORD placed under a curse."[cc] The Bible writers gave Noah his name because of the role he was to play in this story. The name "Noah" is connected with the word "relief" in this passage and is also associated with the phrase "to comfort." But the important thing to recognize about this sentence is that it must be referring to another part of the story. The statement is a prophecy and it makes sense that the prophecy would be fulfilled within the story of Noah. One could consider it fulfilled when Noah improved farming by inventing tilling of the soil. That improvement could have the effect of countering God's curse by forcing the earth to be more productive than it otherwise would have been. That view makes sense, but ignores the Nephilim who, through their consumption of meat, had already achieved true relief from the curse on the earth. So if Lamech is looking at people who had freed themselves from God's curse, his prophecy can be viewed as a longing for the freedom those people had. So Lamech was predicting that Noah would give his people what the Nephilim had. And it was Noah's ability to *compromise* that delivered humankind from the curse God placed on the earth.

This flood story is clearly an original work, even though it is similar to other ancient flood stories. The other stories of Genesis 1 to 11 are also original, even though they seem to bear a more than passing resemblance to other ancient stories. The story of Seth's genealogy is clearly an original work intended to encode a nearly perfect solar calendar. The story of Noah and his ark, together with the story of Creation in Genesis 1, were written to encode and to teach a theory of history. I first recognized this intent when I read the story of Adam and Eve. That story manages to list human needs in the order of their importance in sustaining life. The story begins with breath, moves on to food and water, then to companionship and sex.

As a part of their lesson on world history, the authors use the story of Noah to teach a part of their theory of human origins. That theory would have taken into account the most reliable information the authors had about the history of the human race. They may have heard stories that their ancestors were much smaller, were originally vegetarians, and experienced a dramatic increase in size once they began to eat meat. The writers may have also heard stories about the conflicts between humans of different sizes and abilities. Since our ancestors were thought to be vegetarians, it would be easy to imagine that all animals in the early history of the world ate plants and plants only. This was never true, of course, but one could easily come to this conclusion by extrapolating backward to a "theoretical" beginning. The vegetarian animal world would be the way the world "should" have been. For all these reasons, this story and the history it encodes were never meant to be taken as the absolute literal truth. They were meant to be the best understanding of the world at this time by these authors, and as such, this theory is not different in kind from any modern theory of history.

If you were already familiar with the story of Noah and his ark, you might have noticed that I did not address the issue of how the ark could hold a male and female pair of every animal on earth. The authors tell us that God caused a pair of each kind of animal to come to Noah in his ark. Amazingly, this was not the only time in this collection of stories that all of the world's animals were drawn to one man. In the Garden of Eden story, God also caused every animal on earth to come to Adam so he could name their species. Once you know that there have been two mass animal migrations to two different individuals, you might suspect that the two events are related. They are. But to understand what the animals in Noah's ark represent, we must first understand what they represent in the story of the Garden of Eden.

CHAPTER 5

Genesis 1 to 11

5.1 Sliding Stories

GENESIS 1 TO 11 CONTAINS a series of related and intertwined stories. The first is the story of the creation of heaven and earth. The text of the Bible explicitly refers to the story as a story: Genesis 2:4 says, "Such is the *story* of heaven and earth when they were created."[cf] We know that as a part of the Bible, these chapters of Genesis are a part of a system of education designed to detect and expose those who have insight. So we expect to find that these chapters are primarily educational material.[cg] For the Bible writers, that educational material would be a collection of riddles and puzzles. Fortunately for us, the stories of Genesis 1 to 11 are presented in a form that is now familiar.

We've seen in the diagrams of Chapter 2 that these puzzles use "sliding lines" to great effect. See Figure 2.9, for instance. What's so very interesting about this technique is that the storytellers use it as well. They use "sliding stories." The storytellers produce an overall story by creating story segments. Each segment seems to be mostly unrelated to the other segments. But in reality, each segment tells or retells part of the overall story. The writers use this technique to make the collection of riddles into a puzzle. The reader's task is to recognize that each story is repeating information from the other stories and adding significant information of its own. One has to have the particular insight that the stories are repeated retellings of the same overall story with details added and removed. Others have noted that the stories seem to be repeated, but have not made the next step to understand how deep the repetition goes.[ch] If we represent each "story segment" as a line segment, we can "slide" them one against the other to produce the single overall story.

5.2 Heaven and Earth

The first story is told in Genesis 1:1 to 2:4a. It describes how the heavens and the earth were created out of water and darkness. From our analysis of the philosophy/theology of the Bible writers in Chapter 3, we know that the water and the darkness are the same thing (see Section 3.2.1, "God the Son and Energy"). The story tells us that a wind emanated from God and blew across the surface of the mixture of darkness and water and that the wind was somehow mixed with them both. We also know from our analysis that the wind that emanated from God was the same as the light God created at that time. So in the beginning, there was a light-dark-wind-water mixture.

The Creation story occurs over seven days. On the first day, God separated the light from the darkness. With the equivalence of the light with the wind and the darkness with the water, the separation of the light from the darkness implies the simultaneous separation of the wind from the water. Much of the Creation story follows this pattern of separating two things that are mixed together, or somehow transforming an existing thing into a new thing.

On the second day, God separated the waters below from the waters above and created the sky. Because the waters and the darkness were the same thing, God's action caused the waters below to be separated from the darkness above. The dark waters were left below, and the dark sky was left above. But they remained equivalent, and they are both symbolic of death. On the third day, God separated the waters into seas and the dry land appeared. God then extracted living plants from the living land. On the fourth day, God created the sun, moon, and stars in the first act of original creation so far in this story—they were not separated from a mixture of light and the darkness above. The new lights separated day from night and created the calendar,[ci] so, in effect, God created the calendar on the fourth day. In this way, the authors let us know how important the calendar was to them.

On the fifth day, God extracted the swarming creatures, the birds, and the sea monsters from the seas he had created on the third day. On the sixth day, God extracted creeping animals and beasts from the earth he had exposed on the third day. He then made his second act of original creation: man. God created man to rule over his creation. And in order for man to rule, he was created in the image of God. God created lots of men and women and commanded them to increase in number. Then God gave the very first dietary law: men and women were commanded to eat plants only. In fact, that dietary law was even more restrictive: mankind was allowed to eat only seed-bearing plants (that is, those that produce fruit). God then gave a dietary law for all the other animals: they were to eat only plants as well. In these verses the authors describe their idyllic earth that would be destroyed when men began to eat flesh in the time of Noah.

On the seventh day, God rested from his creative work. The process of creation began with God the Father who, as the Face of God, was and is unapproachable, unknowable, and unchangeable. Out of God the Father blew the wind that we call God the Son, and he created everything that was created. The material world was the object of God's creative work. That material world, just like God the Father, remains unchanged throughout eternity. Anything that moves in the material world moves under the influence of the relationships (or angels) that make up the Son of God. God's rest on the seventh day is symbolic of the material world. The material world is "God at rest."

5.3 Garden of Eden

In Genesis 2:4b to Genesis 3:24, the authors tell us a second story. This story begins with a marker that relates it to the Heaven and Earth story above. It then retells that story with expanded details. The text of the Bible says that before God created man, a flow would well up from the ground and water the whole surface of the earth. That statement is an allusion to the earth being covered with water in the Heaven and Earth story and also to the earth being completely covered with water in the story of Noah's ark. The other marker is the wind from God that blows over the surface of the earth to create man from the dust. That wind parallels the wind from God that blew across the dark waters in the Creation story. The wind in this story reveals itself as man, and the wind from God revealed itself as light. The authors then go on to expand on this new Creation story in much more detail.

This story contains an interesting play on words. "Adam" means man and is also very similar to the Hebrew word for "earth." So when the story says that God formed man from the dust of the earth and blew into his nostrils the breath of life, that statement can be understood in a number of ways. We can understand this story better by taking note of the corresponding creation of man in the story above. The two stories tell us something important about mankind. In the Heaven and Earth story, God created man in his image, and he also created mankind, both male and female. Since God is not an animal, being created in God's image is *not* describing the creation of man's body. But when God creates mankind as male and female, he *must* be creating their bodies, since any male or female characteristic is derived from the physical characteristic. So that story is describing both a physical and a spiritual creation. In this story, it is the soul of man that is created when the spirit of God blows across the earth and kicks up dust from the ground. The writers describe this process as God blowing life into man's nostrils. The similarities between the stories should lead us to conclude that the creation of both the soul and the body of man are being described in this story as well. It is easy to think that the use of the word "nostrils" implies the creation of man's body as well as his soul, but it does not. The use of that word is a misdirection. We're going to pass on that misdirection here.

The first thing that God did after he created man (the soul of man) was to plant a garden in a place called Eden. God put the man in the garden to take care of it. The writers describe the Garden of Eden as enclosing an area where a river flows out of the larger Eden and divides into four separate rivers. The "Eden River" divides within the Garden into the Pishon, the Gihon, the Tigris, and the Euphrates rivers. We know where the present-day Tigris and Euphrates rivers are, and we know where they meet. According to this story, the Pishon and Gihon must have met the "Eden River" not too far from where the Tigris and Euphrates meet. Where did the Pishon and Gihon rivers flow? We're not sure, but some theories associate them with ancient rivers that might have flowed through present-day Iran and Kuwait.[cj]

This array of confusing clues implies that we are dealing with a new riddle, and that the Garden of Eden is *not* what it appears to be. The main clue that this is a riddle is that a river flows out of Eden into the garden, where it divides into four major rivers. This simply does not happen. The waters of the Tigris and Euphrates merge into a single river, which then continues to the Persian Gulf. The direction of the flow in the story is the *reverse* of the direction that the real rivers flow. So the rivers described in the text of the Bible are not exactly the physical rivers of the real world but are some kind of reference to them.

If we try to reconstruct the rivers of the Garden of Eden in the real world from what we know, we know that the Tigris and Euphrates rivers meet at a point and create a single "Eden River" that flows directly to the Persian Gulf. We can then attach to the Eden River an imaginary river Pishon and an imaginary river Gihon to produce the diagram shown in Figure 5.1:

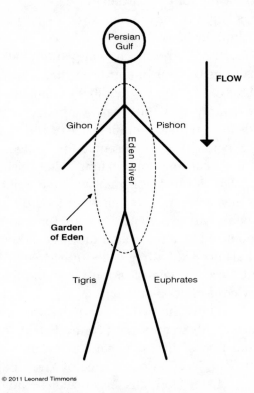

© 2011 Leonard Timmons

Figure 5.1: Garden of Eden. When the rivers that flow through Eden and the Garden of Eden are drawn as described in the Bible, the result is the human form.

So what the Bible writers are describing in this story about Eden and a garden within Eden is the human body. We know that the Pishon and Gihon rivers are the arms of the figure, because they are both *winding* rivers. The way they're described brings to mind the actions one can perform with an arm, that cannot be performed with a leg. We also get the hint that one of the arms is bedecked with jewels. The implication is clear: This story is describing the creation of both the soul and the body of man. The creation of the body of man is told as God planting the Garden of Eden. The authors hint at this when they name the man "Adam" (earth) and create him from dust—a mixture of earth and wind. That dust is more ethereal than the earth and is energized by spirit (symbolized by the wind). That dust is Adam, and Adam is a soul. When God put "Adam" into the Garden of Eden, he put the soul of man into his body. In addition, God made it the job of the soul of man to care for the body of man when he gave Adam the job of caring for the Garden of Eden.

Lots of things begin to make sense after we know that the Garden of Eden is the human body. The word "eden" means "delight" or "pleasure" or "luxury."[ck] And when God planted the Garden in Eden, he caused trees to grow from the ground that were "pleasing to the sight and good for food."[cl] When the Bible writers describe the trees as "pleasing," they imply that the trees symbolize pleasure. Once we know we are dealing with the human body, and that each pleasure the body can experience is represented by a tree growing in the Garden of Eden, we can then understand what this part of the story means: God created the body of man with the ability to experience different pleasures, and man is allowed to indulge in all of the pleasures his body can experience, except for one. That forbidden pleasure is symbolized by a tree called the "tree of knowledge of good and evil." God also caused a tree called the "tree of life" to grow in the Garden, but there is a complication involving the tree of life and the tree of knowledge of good and evil.

When we look at how the two trees are described in the text of the Bible, we are faced with an important contradiction. When the writers first describe the tree of life, it is said to be in the middle of the Garden of Eden. The location of the tree of knowledge of good and evil is not given.[cm] Later God tells Adam that he must not eat of the tree of knowledge of good and evil or he will die.[cn] When Eve talks to the serpent, she says that the commandment she received from God concerns the tree in the middle of the Garden.[co] That tree is the tree of life. She further comments that she is neither to eat of it or touch it or she will die.

This story is implying that the tree of life and the tree of knowledge of good and evil are the same tree, the one tree in the middle of the Garden of Eden. A pleasure in the midst of the human body that is both the tree of life and the tree of knowledge of good and evil is clearly sexual pleasure. This was the pleasure that the newly implanted soul of man was to avoid. When Eve added that she was not even to touch the tree of life, the authors make a clear reference to the taboo in that culture of a woman touching a man's genitals.

After God commanded Adam not to eat of the tree of knowledge of good and evil (God commanded him to avoid sexual pleasure), God attempted to find a companion for Adam among the beasts that were created with mankind on day six. God could look for a companion for Adam among the beasts, because if a relationship did not involve sex, any animal could be a man's companion. So God brought each kind of animal to Adam to see what Adam would call that kind of animal, and Adam named them all. Up to now in this story, God had done the naming. When the authors let Adam name each kind of beast on earth, they repeat the idea that man is made in the image of God, because Adam is performing a task previously reserved to God. We know that it is the *mind* of man that is made in the image of God because naming is a mental activity. The authors use this mechanism to make it clear that the soul of man is the basis for the mind of man.[29] So we now know that the mind of man is like dust—partially spirit and partially flesh. The writers also make it clear that there is no mind in any of the beasts that is equal to the mind of man. The animals were brought to Adam so he could select one that could help him care for his body. (His body is the Garden of Eden, and his mind is what we're calling Adam.) No beast had a mind with an intelligence that would allow it to help Adam care for his body. So to create such a mind, God shut off the mind of Adam by casting him into a deep sleep, and from his soul God created the soul and the mind of Eve. Her mind is equivalent to, but different from, the mind of Adam. She is able to help him care for his body—his personal Garden of Eden.

Since this story is about sex, the introduction of the woman lets the writers address the subject directly. The man and the woman were naked but felt no shame. The ensuing story about the serpent tricking Eve into eating from the tree of knowledge of good and evil is a story of sexual lust. Eve lusted after Adam and convinced him to have sex with her, and in doing so they both partook of sexual pleasure, which was forbidden. The serpent in the Garden is a reference to the male penis as the cause of Eve's lust. To limit its effects, Adam and Eve clothed themselves with fig leaves after they became aware that nakedness could cause lust. The use of fig leaves also brings to mind the fruit of the fig tree, which also resembles the male genitalia. When the writers use the serpent and the fig in this story, they want to make it clear that they are talking about the male sex organs.

What's going on here is that Adam has two commands that seem to contradict one another, but in fact do not. God's commandment to

---

29 A mind is a "living" or energized soul.

Adam is a riddle. God told Adam that he *should* eat from the tree of life that is in the middle of the Garden. Then God commanded Adam *not* to eat of the tree of knowledge of good and evil, which, as far as Adam could tell, was in the middle of the Garden. God created this riddle for Adam and Eve so that they could figure out what the riddle meant and thereby prove that they had insight. Eve failed to have insight into the riddle. Her lack of insight is revealed when she is tricked by the serpent into eating from the wrong tree.

Now that we're sure the tree of knowledge of good and evil is actually sexual pleasure, we are left to wonder what kind of pleasure the tree of life represents—a pleasure that a person could freely indulge in that would ultimately lead to eternal life. The tree of life is not our ability to live through our children by sexual reproduction, though that is what Eve thought. In Section 3.4, we discussed how wisdom's goal was eternal life, and how much wisdom would be necessary to achieve life everlasting. From that point of view, this story implies that the pleasure represented by the tree of life is the pleasure associated with wisdom. A person would have to take pleasure in acting wisely in order to achieve eternal life. We know that Eve is aware of the pleasure associated with wisdom because she says that the fruit of the tree of knowledge of good and evil looked like it would make one wise, but she failed to recognize the riddle. She confused the physical with the spiritual.

God gave Adam and Eve this riddle to allow them to show whether they did or did not have insight. They needed insight to make sense of the knowledge they possessed and to develop enough wisdom to live forever. They showed that they didn't have insight, and without insight they could not acquire that wisdom. In this story, the serpent represents lust, and that lust is shown as affecting Eve's mind adversely. The implication is that her ability to recognize the riddle and figure it out was hampered by her lust. She was tricked by it.

After Adam and Eve had sex, they realized that they were naked. Their shame at being naked indicates a knowledge of guilt for having broken God's law. They realized that there was good and evil in the world and that they themselves were evil. Knowing good from evil immediately imposes on those who know the difference the responsibility to do what is good and to avoid doing what is evil. Before she lusted after Adam, Eve was confident that the benefits of indulging in sexual pleasure far outweighed its potential harm. She saw sex as a tree of life and as a source of wisdom—something that she and Adam should experience. Eve was right on every point. She saw no reason why sexual pleasure should lead to physical death. Yet Eve failed to consider our point above, however, that knowing good from evil imposes on the knower the responsibility to do good and avoid doing evil. She and Adam were unable to avoid doing evil, and that was one of the things that they came to know after they became "like God, knowing good and evil." When the wind from God blew through the leaves of the trees in the Garden, Adam and Eve's inability to avoid doing evil caused them to hide themselves from God's presence.

The story continues with God finding out that Adam and Eve had come to know the difference between good and evil but were unable to do good and avoid doing evil. Their failure to do good and to avoid doing evil is told as God cursing Adam and Eve and driving them from the Garden of Eden into Eden proper. Being driven out is a theme that is repeated as a marker in the stories that follow. The driving of man out of the Garden into a barren Eden represents death or corruption and accomplishes the warning that God issued to Adam that should he eat of the tree of knowledge of good and evil, he would die that very day. Adam and Eve had corrupted themselves.

The authors have God issue a list of curses based on the sin of Adam and Eve. The first curse is issued to the serpent and is used as a marker in subsequent stories. The serpent is made the lowest of all the animals. And this is the authors' way of telling us that sexual desire represented by the male penis is the basest of impulses.

The authors then have God issue a second and very cryptic but interesting curse. God says to the serpent:

> I will put enmity between you and the woman,
> and between your seed and her seed;
> he shall bruise your head,
> and you shall bruise his heel.[cp]

This curse says that there is a permanent state of hostility between the serpent and the woman. A state of hostility also exists between the seed of the serpent and the seed of the woman. While a modern interpretation might lead us to think that the "seed" of the serpent is a man's emission, that is most certainly not what is meant here, since the authors refer to the "seed" of the woman as well. If the "seed" of the man is stored in his sex organs, then it makes sense that the "seed" of the woman is stored in her sex organs as well. So in this instance, the "seed of the man" and the "seed of the woman" must refer to the external male and female genitalia. When the authors refer to the "head" of the serpent, they are making a general reference to the fact that a man's penis bears a resemblance to a serpent. And when they refer to the "heel" of the woman's "seed," they are making a reference to the general similarity between a heel and the appearance of a woman's genital area.[30] The bruising of the serpent's head and the "heel" of the woman's "seed" is a direct reference to the act of having sex, which leaves both organs bruised.[31] This curse is saying that humans are driven to ("shall") have sex.

In the third curse, God says that he will multiply the woman's pain when she gives birth, force her to lust after her husband, and give her husband power over her. If we understand the story simply, at this point in history no woman has ever given birth, so multiplying the pain in childbirth makes no sense. What the authors are implying is that our nonhuman ancestors had pain during childbirth, but that the pain became much, much worse as we became human.

In the fourth curse, God curses the earth for man's transgression, and the earth holds back its abundance. The lack of food makes it extremely hard for people to survive. Ultimately we discover that this curse is related to sex as well. The increase in population caused by sex exhausts the available resources until people have great difficulty feeding themselves. This population explosion is what destroys the Garden of Eden.

After the curses, the story continues with God placing the cherubim and the fiery, ever-turning sword at the entrance to the Garden of Eden. The cherubim and the sword guard the way to the tree of life. This symbology is the authors' way of saying that deep understanding and insight are necessary to experience the pleasure associated with the process of living forever. More importantly, one must find pleasure in having insights, acquiring knowledge, and using one's knowledge and understanding for good. This is the answer to the tree of life riddle. At the time this story was written, the

---

30 A detailed description would be impolite, but the sole of the foot associated with this heel would face forward, and the toes of the foot would be at the navel.

31 My apologies to those of you who have been taught that this verse is a prophecy that predicts the birth of Christ. Clearly that idea was not even close to the intent of the authors.

mind was believed to reside in the belly or the heart. Finding pleasure in having insights would radiate from this area, which is in the middle of the "Garden of Eden."

When we look at this story from a more real-world perspective, we see that it refers to important events in human history. The first men were gardeners: they improved the growing conditions of plants that they valued so that those plants fared better than other plants for which they had no use. We see that early on, women and men lived together as equals with a common purpose. Unlike the females of most mammals, women became sexually available at all times. The increase in sexual activity led to disease, until some form of monogamy was instituted. (A man leaves his mother and father, and he and his wife become one flesh.) The size of the human head increased and thereby increased the pain of childbearing. Using grains to make bread produced one of the first processed human foods. And people first clothed themselves with plants and later with animal skins.

The other real-world issue is whether a geographical Garden of Eden really existed. If we overlay our diagram in Figure 5.1 with the real Tigris and Euphrates rivers, the Garden of Eden would be the area from just north and west of the current-day city of Basra to the waters of the Persian Gulf. Man did not originate within the Garden. It's where he developed his soul or a mind that could be like the mind of God. The writers tell us that the Garden was in the east, so they would have been somewhere west of the Euphrates, maybe near ancient Israel.

This story does not say that only one man and one woman were the first human beings. Many people would have been in the Garden of Eden. God created them in Genesis 1:27–28. One family could have contained a patriarch who would have symbolized the man, Adam, and his wife would be symbolic of the woman, Eve. Their children could have married individuals from the other unnamed families of the Garden of Eden.

5.4 Abel and Cain

The third story is told in Genesis 4:1–16. This story is about money and worship. It is also an expansion of the Garden of Eden story. Two obvious markers in this story correlate with that story. The first marker is that Adam and Eve have sex again. Their sex act is the centerpiece of the story above. The second marker occurs at the end of this story when Cain is driven from the presence of God. Cain's expulsion corresponds to God driving Adam and Eve from the Garden of Eden.

The markers imply that this story corresponds to the part of the Garden of Eden story that occurs after Adam and Eve have sex, but before God drives them from the Garden. In this story, Eve names her first child Cain, which in Hebrew is similar to the word for "gain" or income or what we would now think of as money.[cq] Cain's name implies that this story is about gain and the lust for gain. Today we would say that this story is about money and the love of money. In the Garden of Eden story, trees in the Garden represent the pleasures of the body. In this story, we are introduced to the pleasure associated with wealth.

After they had sex again, Adam and Eve had a second child they named Abel. It is important to note that in the text of the story, Abel is mentioned before Cain, even though he is the second son.[32] Abel got gain by raising animals, and Cain got gain by raising plants. When Cain killed Abel, he disgraced himself and lost his inheritance. His disgrace means that Abel comes before him in the line of inheritance. So when they are both the subject of discussion, Abel comes before him and is listed first. More on this later.

It came to pass that Abel and Cain each offered some of the fruits of their labors as a sacrifice to God. Abel offered the very best that he had produced as a sacrifice. Cain sacrificed something that he had on hand. God accepted Abel's sacrifice and rejected Cain's sacrifice. Cain became upset with *Abel*, not with God. He asked Abel to meet him in a field, where he fell upon his brother and killed him. Falling upon his brother brings to mind the Nephilim, since their name derives from their practice of falling upon their prey. Cain actually hunted his brother like the Nephilim would one day hunt their prey.

The sacrifices that Abel and Cain offered to God were not intended to feed those making the sacrifices (the most common use of sacrifice). Instead, fire completely consumed the sacrifices. The item that was sacrificed would have been a total loss. Sacrifice is an act of worship, so we need to know what worship is, in order to understand what's going on in this story. Worship is a certain kind of teaching. Worship teaches about wisdom and everlasting life.

---

32 Genesis 4:2 is the only reference where this is explicit.

Everlasting life can be achieved through an infinite series of wise acts that are good because they preserve the life of the actor. Foolish acts are evil because they limit the ability of the actor to live. Worship is about demonstrating this theology in words and actions. A single evil act, or an act of worship that does not measure up, can lead to death. So a worship ceremony is the symbolic reciting of this theology in words and actions. Worship is not a means by which one can achieve everlasting life, except in its role of teaching others. And in that role, the symbology can be confusing, since the worship lesson would be taught in a way that would make it understandable only to those with insight. In particular, worship is about performing an act, or a series of acts, that teaches life-preserving behavior to those who have insight.

In this case, the point of Abel and Cain's acts of worship was to produce evidence that they could divorce themselves from the pleasure of wealth. They would provide that evidence by destroying something of value to each of them. Wealth represents stored pleasure, because things of value can be converted into things that are more immediately pleasurable. The lust after wealth or the love of money is a thing that leads to death and must be avoided. The problem with Cain's worship was that he offered some of his gain, but he did not offer the most perfect (and therefore the most valuable) of his crops. He held back. Cain was of the opinion that he could be good (and thus live forever) by giving God an offering that was flawed. It was too difficult for Cain to give up the stored pleasure in the valuable food that he would have destroyed. On the other hand, Abel offered the most precious of his lambs. God accepted Abel's offering as being symbolically correct, and his worship also proved that he did not love wealth. Cain's act of worship did not measure up because it did not teach the lesson of the theology—the need to be perfect to live forever. Abel's act of worship measured up, because in offering the most perfect of his lambs, his act was precisely symbolic of the theology. The perfect lamb symbolizes a series of acts, all of which are wise, and none of which are foolish.

Cain's reaction to God's unwillingness to accept his sacrifice was to become angry and sullen. God then says,

> "Why are you distressed,
> And why is your face fallen?
> Surely, if you do right,

> There is uplift.
> But if you do not do right
> Sin couches at the door;
> Its urge is toward you,
> Yet you can be its master."[cr]

These verses are difficult to understand because of the term "couching at the door," which seems to be an idiom from this time.[33] If we try to deconstruct the idiom, "to couch" would be a good way of rewriting "couching." A couch is what we would call a bed. So "to couch" would be to lie on a bed. Why would someone lie on a bed near a door? The passage gives us the information that allows us to figure out why. Whoever is lying on that bed has an urge toward someone on the street, and that someone is Cain. For his part, Cain can give-in to that person's urge, or he can avoid it. The author is clearly describing prostitution. Presumably, at this time a harlot would lie on her bed near a doorway to entice her clients.[34] She would be "couching at the door."

This is now the third marker in this story, and it refers to the sexual sin of Adam and Eve in the Garden of Eden and the "sexual" sin of the daughters of men in the story of Noah's ark. So "couching at the door" is our marker to those events in those stories. The pleasure under discussion is again sex, but sexual pleasure is being obtained in exchange for money. Wealth, as money or other things of value, is the subject of this story, and prostitution merges sin, wealth, and pleasure into a single act. The sin under discussion in this story is making wealth your ultimate goal in life, and that is essentially the same as making pleasure your ultimate goal in life. When you make wealth your ultimate goal, you become the prostitute "couching at the door" for money. As the prostitute, you are willing to do anything, even evil things, to obtain wealth. If wealth is your ultimate goal, you also become the client of the prostitute, who is so focused on obtaining

---

33 Other versions of the Bible may say "crouching at the door" or something similar. Now, it is certainly true that one definition of "to couch" is for something or someone to lie in ambush. The use of the word "crouch" brings to mind a powerful animal that might see you as a meal. You certainly could get that meaning from this cryptic passage. We do not know why a door has to be involved in any kind of crouching, however. Yet the door is there in the story. That's why I think the JPS Tanakh translation is correct to use the word "couching," and I explain in the text why I think the more common definition, to lie on a piece of furniture, is appropriate.

34 She could even crouch on her couch to entice her clients.

pleasure that he is willing to do all kinds of evil to obtain that pleasure. You must master both your urge to possess pleasure (the stored pleasure that wealth represents) and wealth's urge to possess you. Wealth is a tool; it is not and must not be an end in itself. Cain prostituted himself to obtain wealth and was unwilling to part with his wealth unless he was going to receive something in return.

The wealth angle becomes doubly apparent when the text seems to imply that Cain took possession of Abel's flock after Abel's death. Not only was Cain unwilling to waste his own wealth on sacrifice, his killing of Abel prevented Abel from wasting *his* wealth on sacrifice. If that wealth came into Cain's possession after Abel's death, then by killing Abel, Cain also stole Abel's flock. It certainly would appear that Cain killed Abel to take his wealth, which, from Cain's point of view, Abel was wasting.

The people living near Abel and Cain would wrongly conclude that Cain murdered Abel for his wealth, and they would believe that Abel's death should be avenged. In order to avoid this fate, Cain escaped to an area where he was less well known. The authors tell this as God banishing Cain to the land of Nod east of Eden, and we are left to wonder why God would prevent Cain from suffering the consequences of his murderous actions. The answer is that the conflict between Abel and Cain was about their freedom to worship as each saw fit. The authors want us to understand this, and they have God put a mark on Cain so that he will not be killed for the wrong reason.

It becomes our task to figure out that religious intolerance was responsible for the death of Abel. Anyone who found Cain and desired to avenge Abel's death would first have to understand that Abel died at the hand of a man who disagreed with the way he worshiped. Once Abel was dead, Cain took that opportunity to steal Abel's possessions, but that was not the motive behind the murder.

One of the real-world events that defines us as human beings is the rise of religious intolerance. That intolerance creates conflicts that ultimately help break groups of people apart and disperse them across the earth. This story tells us about the first violent act of one man against another, and that act was an act of religious intolerance.

Another real-world event that the authors are referring to in this story is the rise of violence early in human history, where one group attacked another to take that group's wealth. This activity was fundamental to the process of humans evolving into who we have become. Groups who no longer lived off the land but who survived by attacking and exploiting settled peoples became another class of wanderers. Ultimately, their lifestyle would result in the culture alluded to in Cain's genealogy, a culture of wandering warriors.

5.5 Calendar and Flood

The fourth story in this sequence of stories is told from Genesis 4:17 to Genesis 8:14. This is the story we first discussed, the story of Cain's genealogy. That genealogy introduced us to the calendrical puzzle hidden in the genealogy of Seth. Cain's genealogy was designed to give us the clues we needed to decode the calendar in Seth's genealogy. Another interesting thing about Cain's genealogy is that the women are named and, as a consequence, exactly seven people are mentioned.

The genealogy of Seth leads us directly to the calendar, a beautiful encoding of a nearly perfect solar calendar.

Woven into the genealogy of Seth is the story of the Flood. The Flood story tells us that everyone and everything on earth was vegetarian at some point in the past. It also tells us that human populations are prone to population explosions and crashes, and those explosions were first caused by the consumption of flesh. It also tells us that laws controlling the consumption of flesh and the rules governing sacrifice help prevent those population explosions and the subsequent crashes.

The story of the Flood also tells us about the first war and the first siege. We find that early fortifications were created on natural or man-made mounds, a technique that may have evolved into the building of higher and more elaborate mounds that we recognize as ziggurats. Eventually, those ziggurats became as large and complex as the Tower of Babel. The authors provided the details of the siege to give us the information we needed to decode the internal workings of the calendar.

At this point, we've covered the puzzles and the riddles from Creation up to the end of the Flood. They're mostly in context and understandable. Earlier, we noted that the Creation story was being retold in the middle of the Flood story when the earth was covered with water and darkness. As the stories continue, we will find that the retelling of the Creation story begins the retelling of the overall story.

CHAPTER 6

The Story of Shem

6.1 Shem's Birth

AFTER THE FLOOD, NOAH'S SONS become important, and the genealogy of Shem takes over. The text of the Bible says that Noah's sons were born in his 500th year, and the Flood occurred in Noah's 600th year. Since Noah's sons were not triplets, we would expect that the oldest son would have been born in year 500, and the others would have been born in later years. The text always refers to Noah's three sons in the order: Shem, Ham, and Japheth. Repeatedly listing the sons in this order implies that Shem was the oldest, Ham was the second oldest, and Japheth was the youngest.

The text also says that Shem was 100 years old two years after the Flood, when Noah was 602 years old. Shem, therefore, had to be born in Noah's 502nd year. Since the text says that Noah's first son was born in his 500th year, Shem could not have been Noah's firstborn son even though he is always listed first.

The text also says that Shem was Japheth's older brother, so Japheth had to be born after Noah was over 502 years old. These facts might lead you to believe that Ham was the oldest son even though he is listed second. But Genesis 9:24 seems to say that Ham was Noah's youngest son. The confusion here is massive. Does the reference to "Noah's youngest son" in verse 24 refer to Ham or Canaan? It could be referring to Canaan, but Canaan is not Noah's son. A straightforward reading of the text seems to say that none of Noah's sons were born in his 500th year.

The confusion is not over. In Genesis 9:18, the story of Noah's drunkenness and the condemnation of Ham's son Canaan begins. The story is one of the strangest yet:

> The sons of Noah who came out of the ark were Shem, Ham, and Japheth—Ham being the father of Canaan. These three were the sons of Noah, and from these the whole world branched out.

Noah, the tiller of the soil, was the first to plant a vineyard. He drank of the wine and became drunk, and he uncovered himself within his tent. Ham, the father of Canaan, saw his father's nakedness and told his two brothers outside. But Shem and Japheth took a cloth, placed it against both their backs and, walking backward, they covered their father's nakedness; their faces were turned the other way, so that they did not see their father's nakedness. When Noah woke up from his wine and learned what his youngest son had done to him, he said,

> "Cursed be Canaan;
> The lowest of slaves
> Shall he be to his brothers."

And he said,

> "Blessed be the LORD,
> The God of Shem;
> Let Canaan be a slave to them.
> May God enlarge Japheth,
> And let him dwell in the tents of Shem;
> And let Canaan be a slave to them."[cs]

It is not at all clear what is going on in this story. Noah gets drunk and naked, and Ham sees him. At this time in the history of this culture, seeing one's parent naked was completely and absolutely taboo, and it was to be avoided at all costs. Noah issued a curse when he found out about Ham seeing him naked, but he cursed his grandson Canaan. Why? What did Canaan do? Who is Noah's eldest son? What does this story have to do with the way people branched out over the earth after the Flood?

This story occurs immediately after the Flood, so let's look at it in that context. The Flood reproduces the Creation story, except that at the end of the Creation story all animals ate plants. At the end of the story of the Flood, animals could eat anything. Immediately after the Creation story, the Bible writers relate the story of the Garden of Eden. That story is about the introduction of sexual lust into the world.

In the story of the Garden of Eden, the Garden is the human body, and Eve believed the tree in the midst of that garden represented the male sex organ. (See Section 5.3, "Garden of Eden," for more details.) The fruit of that tree was sexual pleasure, and God commanded Adam and Eve not to indulge in that pleasure. The serpent represents the temptation of lust. The serpent is a trickster who spoke to the woman and told her that experiencing sexual pleasure would make her like God (because she would come to know good from evil), and she believed him (he was right). Her lust caused her to convince her husband to have sex. They became aware of their nakedness (symbolic of evil), and they hid themselves. Once they knew the difference between good and evil and realized that they were evil, they avoided the presence of God. God found them, clothed them, cursed them, and drove them out of the Garden of Eden.

This story of Noah's drunkenness is a retelling of the story of the Garden of Eden. Canaan is the serpent. Ham is Eve. Noah's tent is the tree of knowledge of good and evil. Shem and Japheth represent God. Noah represents both Adam and the knowledge of good and evil.

What happens in this story is that Canaan is somehow aware that his *grandfather* Noah is drunk and naked in his tent. Canaan tricks his *father* Ham into entering Noah's tent so that Ham will commit the unthinkable act of seeing his father naked. In the story of the Garden of Eden, eating the fruit of the tree of knowledge of good and evil (Adam and Eve having sex) is the equivalent sin. Once Ham commits this act, he is distraught and confesses his shame to his brothers Shem and Japheth. In the Garden of Eden story, the equivalent act is Adam and Eve confessing to God that they had eaten the fruit of the tree of knowledge of good and evil (had sex).

We know that Ham was tricked into entering Noah's tent, because had he entered the tent of his own volition, he could easily have kept the violation of the taboo secret and left. But he couldn't keep the violation secret with Canaan aware and presumably laughing about the situation he put his father in. Shem and Japheth then cover their father to prevent anyone else from being victimized by Canaan's childish trickery. There is no other reason to cover a naked man whose nakedness is already hidden by his tent. In the story of the Garden of Eden, the equivalent act is God clothing Adam and Eve.

When Noah recovered from his drunkenness, someone told him what Canaan did to him and to his son Ham. At that point Noah represents the knowledge of good and evil. In the Garden of Eden story, after Adam and Eve ate the fruit of the tree, their eyes were opened, and they knew that they were naked. When Noah was drunk and naked, he did not know that he was naked.

Furthermore, when the text says that Noah was informed about what his youngest son did to him, it is speaking of Canaan, his grandson, not Ham. Ham is, in fact, Noah's eldest son. Ham was born in Noah's 500ᵗʰ year. Because he saw Noah naked, he lost his inheritance and is listed as the second son in every reference. The birthright then goes through Shem. The same thing happened to Cain. Though he is the firstborn son, in any list, Abel would be listed before Cain. Cain disgraced himself when he killed his brother Abel, and Ham disgraced himself when he saw his father's nakedness. This transgression by Ham ultimately broke Noah's family apart and planted the seeds that caused Noah's descendants to scatter across the earth.

The curse imposed on the serpent in the Garden of Eden story was that it should crawl on its belly forever. It is made the lowest of creatures. When Canaan is cursed, he is made a slave to slaves, the lowest of the low. The curses are the same.

6.2 Branched Out

Genesis 10 and 11 continue to tell the segments of the overall story, but the segments are out of order. The story of Babel is told together with the genealogy of Shem in Genesis 11, while the detailed genealogy of Noah and his sons is told in Genesis 10. We will explore the story of Babel next. It tells us how diverse human languages were created and how they caused people to branch out from a common homeland.

My analysis above showed that the story of Noah's drunkenness is a retelling of the story of the Garden of Eden. In the story of the city of Babel, the earth's entire population lives in a single city, a clear reference to the Garden of Eden. The story of Babel ends with the dispersal of all the people from the city, just as God drives Adam and Eve (symbolizing earth's entire population) from the Garden of Eden. So one might reasonably conjecture that the story of Babel is also a retelling of a part of the Garden of Eden story.

Here is the text of the story, from Genesis 11:1–9.

> Everyone on earth had the same language and the same words. And as they migrated from the east, they came upon a

valley in the land of Shinar and settled there. They said to one another, "Come, let us make bricks and burn them hard."—Brick served them as stone, and bitumen served them as mortar.—And they said, "Come, let us build us a city, and a tower with its top in the sky, to make a name for ourselves; else we shall be scattered all over the world." The LORD came down to look at the city and tower that man had built, and the LORD said, "If, as one people with one language for all, this is how they have begun to act, then nothing that they propose to do will be out of their reach. Let us, then, go down and confound their speech there, so that they shall not understand one another's speech." Thus the LORD scattered them from there over the face of the whole earth; and they stopped building the city. That is why it was called Babel, because there the LORD confounded the speech of the whole earth; and from there the LORD scattered them over the face of the whole earth.ᶜᵗ

The context of this story is that the city of Babel was a real place established by someone in Ham's lineage, either Ham himself, his son Cush, or Cush's son Nimrod. The normal line of inheritance would have flowed through these individuals had Ham not lost his inheritance. Within the city was everyone who was alive. That includes Shem and his progeny. Canaan was also there. The real-world context is that people were moving from a nomadic lifestyle to a settled lifestyle. The place where this happened for the first time might be called a garden, or it could be called a city, but it would be both.

What is skipped in this retelling is an analog to the trickery of the serpent or the trickery of Canaan. Also missing is the prohibition of eating the forbidden fruit. In fact, everything that happens in this story seems to occur after Adam's first sin. So the events in this story correspond to the events in the Garden of Eden story after God has cursed Adam and Eve and clothed them. The cursing of Adam and Eve corresponds to the moment in the story of the drunken Noah after Noah curses Canaan. It thus becomes clear that the story of Babel is a direct continuation of the drunken Noah story. It tells us what it means when God drives Adam and Eve from the Garden to prevent them from eating the fruit of the tree of life.

The hard part of the analogy between these two stories is finding in the Babel story the everlasting life mentioned in the Garden of Eden story. Actually, it's fairly obvious once you think about it. The *name* that the people in the city sought to make for themselves was supposed to be an everlasting name, and in that way they intended to achieve immortality by being remembered for all time.

Now that we know the story of the Garden of Eden is the same as the stories of the drunken Noah and the city of Babel combined, we can figure out what the storytellers are saying overall. The centerpiece of these two stories is the knowledge that good and evil exist. Before eating the fruit of the tree of knowledge of good and evil, Adam and Eve were in a state similar to Noah's drunkenness. After Ham saw his father naked, he was ashamed. After Adam and Eve ate of the fruit of the tree of good and evil, they were also ashamed. There was good and there was evil, and both of them became aware that they themselves were not good.

We discussed good and evil in Chapter 3, so we know that knowing the difference between good and evil has consequences. Eternity is implied. Everlasting life is a possibility, and that potential life exists as a consequence of performing an infinite sequence of good acts. Those acts are good because they keep the person who does them alive. And knowing whether an act is good or evil requires the ability to predict the consequences of that act into the indefinite future. So being faced with the task of actually doing what it takes to live forever is a direct consequence of knowing good from evil.

The people in the city of Babel thought they should create an everlasting name for themselves, and they believed their name would be everlasting life for them. They decided to build a very tall tower that would be a wonder in the world. Each person working on the tower would be working to achieve everlasting life for himself. For this reason, each person would have to be *absolutely certain* (have faith) that his actions would lead to everlasting life.

Let's say that you were one of the Babylonians, and you were sure that an everlasting name was not the same as everlasting life. Or let's say you were sure that building a tower would not create an everlasting name for its builders. Or let's suppose you were convinced that doing something else would lead to everlasting life. From your perspective, deciding to work on the tower would be a *life and death decision*. You would feel that you *had to* follow your own path to everlasting life, and you would disagree intensely with the idea that you should build a tall tower.

If the authorities in the city of Babel taxed you to build a massive tower that you considered a complete waste of time and money, you might feel the need to hide your wealth to avoid those taxes. A simple way to hide your wealth in a market economy would be to use a different naming scheme to refer to things of value. Anyone who did not understand your scheme would be unable to determine how much you owned. In order to determine the value of your possessions, a taxing authority would have to make its own records, and the name confusion would be a significant barrier to any kind of taxation. In order to make a deal involving valuable objects that were not present, one would have to translate between family groups who used different naming schemes. As the taxing authority discovered the meaning of the schemes in use, a "naming race" would develop in which new naming schemes would be created as the taxing authority discovered the meaning of the old schemes. Ultimately you could only communicate with those to whom you revealed your nomenclature, collecting taxes would become nearly impossible and massively unfair, and any collective action would suffer. Not only would it be difficult to build the massive tower, other city services would suffer as well. The city would have to make it impossible to avoid taxation by using a rule that was unavoidable, but also would most certainly be massively unfair.

At this point, you would have to leave the city if you did not want to participate in building a massive tower to create an everlasting name. Others would certainly leave for fairness reasons. So it is possible that this story is an extrapolation back to the very first tax revolt, and that would provide a bit of historicity to the story. Even though the story might detail how a tax revolt led to the creation of diverse languages, the centerpiece of the story is, once again, religious freedom. Religious freedom is the freedom to seek everlasting life in a manner you think will ultimately succeed. You would be driven from the city by your need to act in your own best interest and to avoid acting contrary to your best interests. That need underlies everything that occurs in this story.

The seeking of eternal life is religion. The seeking of eternal life by groups of people is corporate religion. When groups or individuals insist that others follow their religion, they are being intolerant of the religion of others. When you believe that you know what it takes for others to live forever and insist that they follow the path you prescribe, you are practicing religious intolerance. The disagreements thus produced tend to break groups apart, and can become intense enough to result in violence. Cain's disagreement with Abel is an example of this kind of violent intolerance. (Their religious intolerance was based on the proper way to worship.) When the groups break apart, the smaller groups that result can eventually follow the same path and break apart themselves.

Since this story is the same as the Garden of Eden story, we should be able to figure out what it means for Adam and Eve to be driven from the garden. The same thing that happened in Babel must have happened in the Garden of Eden. Adam represented all men, and Eve represented all women, so there was a group of people living in the garden. God created them in Genesis 1:27–28. After they gained the knowledge of good and evil, the group decided that they had to seek eternal life as a single group—that is, they set up a corporate religion. That corporate religion produced religious intolerance within the group, and it caused them to split into smaller groups whose conflicts over religion drove them out of the garden, a place that was made a no man's land by the burden of religious intolerance, just as Babel was.

The story of Abel and Cain that follows the Garden of Eden story is also about religious intolerance. The conflict that arises between the two brothers parallels the conflict that arises in this story concerning which path leads to everlasting life. So the story of Abel and Cain enlarges the story of God driving Adam and Eve from the Garden of Eden, and it also fills in the story of God driving people out of Babel. The missing portion of that story involves an argument between Abel and Cain about what kind of worship is required to achieve everlasting life. Both Abel and Cain seem to have believed there was a single path to everlasting life. And they seemed to believe that the path was through their own form of worship. They did not recognize that worship was symbolic of a theological principle, so Abel's position would have been that Cain had to worship as he did. Cain would believe that Abel had to worship as he did. They came to blows over this issue, and the result was Abel's death.

The fact that they were both wrong about how to achieve eternal life directly corresponds to the Noah's ark/Babel story, in which the people of Babel disagreed over how to achieve eternal life and had to get away from one another because of their religious intolerance. The inhabitants of Babel were all as wrong as Abel and Cain. If Abel and Cain had each allowed the other to seek eternal life as he saw fit, their fight need not have happened; Abel did not have to die, and Cain did not have to be banished. But Cain made three errors. He was wrong about the acceptability of his worship, wrong about whether worship led to eternal life, and wrong about whether everyone had to worship the way he did. Abel was wrong about whether worship led to eternal life and wrong about whether everyone had to worship his way.

The story of Abel and Cain tells us that the magnitude of the conflict between the groups within the Garden of Eden and the groups within Babel may have been intense enough to lead to physical violence. The intensity of the need to achieve eternal life defines us as not just man, but the "son of man."[35] Man as originally created was not much different than the animals. We were in a state of ignorance that was not different from a state of intoxication. But the son of man, the man produced by the knowledge of good and evil, is aware of the concept of eternal life and fights to achieve it.

Something that is clearly implied in this story is that Babel or Babylon is, in fact, the real-world Garden of Eden. Throughout the Bible, Babylon is spoken of as a city of evil. If building a massive tower would not have given its builders everlasting life, then building it was, in fact, a great evil, and the city was full of evil. Their tower was a false tree of life. The same would have been true of the Garden of Eden, since those in the garden would have produced misguided corporate attempts to achieve everlasting life that had to fail.

At the end of the story of the Garden of Eden, God placed a group of cherubs (cherubim) at the entrance to the garden with a single, fiery, ever-turning sword, and together they "guard the way" to the tree of life. One might think that "guarding the way" means preventing someone from taking a path. However, when on a path filled with dangers, as the path to everlasting life is, one might need the cherubs, the fire, and the sword to reach that destination.

The cherub and the sword are symbolic. A cherub is a "higher level" angelic creature. Angels are the relationships that describe and control other relationships. Other higher and higher level relationships lead to the Ultimate Meta-knowledge Relationship (UMkR) at the highest level. The collection of all the relationships and meta-relationships that point to the UMkR is the Son of God who energizes the world. A very high-level meta-relationship, or a

---

35 These authors frequently use the phrase "son of" to mean "successor to" or "derived from."

collection of them, can be called a cherub, and knowing those relationships is what will allow us in any situation to predict the future well enough to do the right thing. That series of righteous acts is what will allow each of us to live forever. The other symbol is the fiery sword. It also guards the way to everlasting life or the tree of life. It is a symbol for the UMkR, and there is only one.

The symbology is complete. The world itself is the garden. The cherubs are the relationships that describe and control the world. The fiery, ever-turning sword is the UMkR. The collection of it all is God. The UMkR (fiery sword) and the world (Garden of Eden) are God the Father. The cherubim as a group are God the Son. You need them all to live forever.

6.3 Shem's Genealogy

Shem's genealogy is similar to Seth's, but the numbers are doublets instead of triplets. The final age of each patriarch is not included in the text, except for Terah. The text of the genealogy from Genesis 11 is included below:

> This is the line of Shem. Shem was 100 years old when he begot Arpachsad, two years after the Flood. After the birth of Arpachsad, Shem lived 500 years and begot sons and daughters.
>
> When Arpachsad had lived 35 years, he begot Shelah. After the birth of Shelah, Arpachsad lived 403 years and begot sons and daughters.
>
> When Shelah had lived 30 years, he begot Eber. After the birth of Eber, Shelah lived 403 years and begot sons and daughters.
>
> When Eber had lived 34 years, he begot Peleg. After the birth of Peleg, Eber lived 430 years and begot sons and daughters.
>
> When Peleg had lived 30 years, he begot Reu. After the birth of Reu, Peleg lived 209 years and begot sons and daughters.
>
> When Reu had lived 32 years, he begot Serug. After the birth of Serug, Reu lived 207 years and begot sons and daughters.

When Serug had lived 30 years, he begot Nahor. After the birth of Nahor, Serug lived 200 years and begot sons and daughters.

When Nahor had lived 29 years, he begot Terah. After the birth of Terah, Nahor lived 119 years and begot sons and daughters.

When Terah had lived 70 years, he begot Abram, Nahor, and Haran.

[The genealogy of Terah.]

The days of Terah came to 205 years; and Terah died in [the city of] Haran.[cu]

When we look at the numbers in this genealogy, it is hard not to notice that Arpachsad and Shelah lived 403 years after the birth of their sons, that Eber lived 34 years before he fathered Peleg, and Eber lived a total of 430 years. The authors seem to be using the numbers 4 and 3 in interesting combinations. Later, the numbers 209, 29, and 119 seem to indicate the authors are toying with the numbers 2 and 9. While these features might be a desired result for the puzzle builders, they do not help us move toward the solution to this puzzle. They are a misdirection. They are intended to lead the person who might attempt to solve the puzzle of this genealogy astray.

When we plot the genealogy of Shem on the diagram of the genealogy of Seth, the last of this new set of patriarchs to die is Eber, and he dies 181 years after the death of Noah. This is our clue to the solution of the puzzle of the genealogy of Shem. See Figure 6.1.

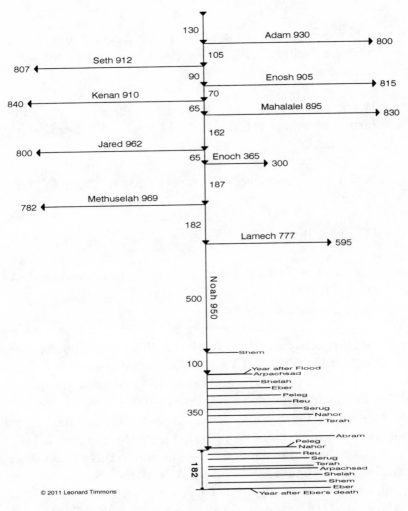

Figure 6.1: Genealogy of Shem. When Shem's genealogy is plotted together with Seth's, the time line is extended 182 years—half of a Divisible Year in days.

Since 181 is so close to 182, we can be fairly sure that the authors are pointing us to the 182nd year after the death of Noah. That year is a hidden reference point for the puzzle. To generate this hidden reference point, the puzzle authors use the same "sliding lines" technique that they used in the earlier puzzles. Remember that the 350-year period at the end of Noah's life was divided into two periods by Hidden Reference Point 2 (see Figure 2.13). The first part of the 350-year period is 182 = 2×91 years, and the second part of the 350-year period is 168 = 2×84 years. If we "slide" the 350-year period down by 182 years, a 182, 168, 182 sequence is generated. The region defined by these numbers and the 98 years to the birth of Shem are the canvas on which the puzzle authors designed this last puzzle in this sequence of puzzles. See Figure 6.2.

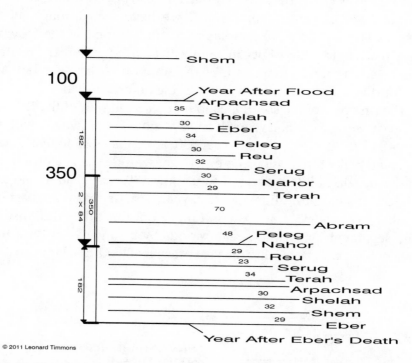

© 2011 Leonard Timmons

Figure 6.2: Time Line Extension. When Shem's genealogy is plotted with Seth's genealogy, the 350-year period at the end of Noah's life is "slid" down 182 years to create the hidden reference point one year after Eber's death.

As in the genealogy of Seth, the deaths of the patriarchs are important, so they are plotted as well. The death of Abram is not plotted, because it is not within the text. Abram's life and genealogy are not a part of this puzzle. As you can see from Figure 6.2, Peleg is the first to die, and he dies one year before Nahor dies. Also, the year after the Flood is one year after the beginning of the Flood. Technically the Flood is not over, since the story of the Flood covers a little more than one year. Arpachsad is born one year later, two years after the beginning of the Flood. Note also that Shem was born in Noah's 502nd year, two years after the birth of Ham. Also note that 13 years separate the death of Terah and the death of Arpachsad. The number 13 is not shown on the diagram because of space restrictions.

6.4 Shem Solved

The solution to this puzzle involves finding the basis for the solution and filling in the particulars. We have already seen the basis of the solution, and it's the "sliding lines" we have seen through all the puzzles above. The next component of the solution is the "toying" with rearranging the digits in numbers to produce other "related" numbers. The third component of the solution is the calendar. And the fourth component of the solution is the number seven.

The first part of the solution is to note that the sum of the 182, 168, 182-year sequence is 532 years. It's not obvious, but if we toy with the digits in the 350-year span, we get 530, which is close to 532. And if we could make the relationship exact, that would be a *very* nice result. So the puzzle builders decided to use the 530-year span as the next "slider." It's easy to "slide" two 530-year periods two years past one another to get a 532-year span, but that's not what they did. Instead, they separated the two 530-year periods by a single year, for a total span of 531 years beginning the year after the Flood. See Figure 6.3.

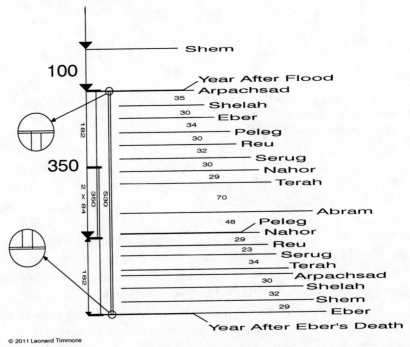

© 2011 Leonard Timmons

Figure 6.3: 350 and 530. The puzzle authors fix the birth of Arpachsad and the death of Eber by inserting dual 530-year periods offset by one year beginning one year after the Flood.

When the puzzle builders "slide" the 530 years by one year, they fix the death of Eber at 181 years after the death of Noah, and they also fix the birth of Arpachsad two years after the Flood. The four lines represented by the year after the Flood and the birth of Arpachsad, together with the year after Eber's death and the death of Eber, are the reference points the puzzle builders used for fixing the births and the deaths of the remaining patriarchs.

The next part of the solution involves three sliders that are all 369 years long—the length of a Leap Year in days. See Figure 6.4.

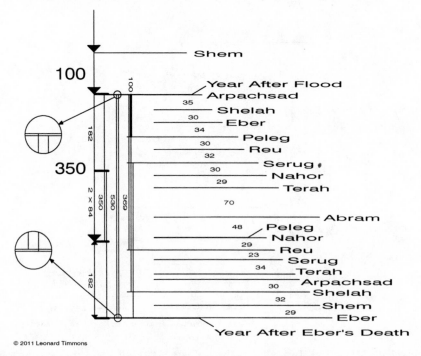

© 2011 Leonard Timmons

Figure 6.4: 369. The length of the Leap Year, 369 days, is used in three separate "sliders" to fix the births of Peleg and Serug and the deaths of Reu and Shelah.

The reference points for the 369-year periods are the year after Eber's death and the year after the Flood. The leftmost slider of this group has one end fixed at the year after the Flood, and the other end sets the death of Reu 369 years later. The third slider has one end fixed at the year after Eber's death, and the other end sets the birth of Serug 369 years earlier. The second 369-year slider has one end that begins 100 years after the year after the Flood at the birth of Peleg. At the other end of the slider, 369 years after the birth of Peleg, Shelah dies. The three 369-year sliders, as a feature of this puzzle, provide additional confirmation that our solution to the original genealogy puzzle was the correct one. It does this by using the number 369, which was the length of the Leap Year in the puzzle of Genesis 5.

The next part of the solution involves two 390-year sliders within the 630-year span from the birth of Shem to the year after Eber's death. See Figure 6.5.

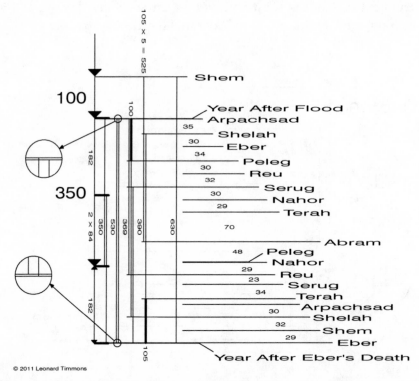

© 2011 Leonard Timmons

Figure 6.5: 390 and 630. Two 390-year "sliders" are used to fix the births of Shelah and Abram and the death of Terah. The sliders are part of a 630 = 6×105-year span.

The reference points for this section of the puzzle are the year after Eber's death and the birth of Shem. The number of years between the year after Eber's death and the birth of Shem is 630. These years divide easily into six 105-year periods. Note that 105-year periods are used as they were in Figure 2.14 and Figure 2.15. The birth of Abram is set to 390 years after the birth of Shem. The death of Terah is set 105 years before the year after Eber's death. And 390 years before the death of Terah, Shelah is born. The two sliders sweep out a period of 5×105 = 525 years.

The next part of the puzzle involves three 309-year sliders. See Figure 6.6.

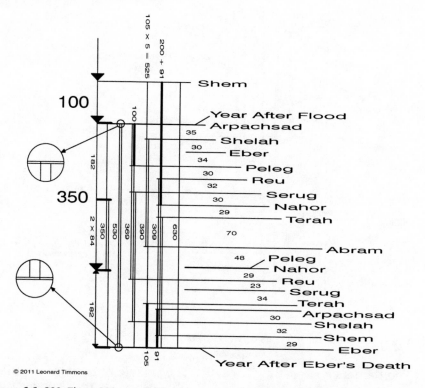

© 2011 Leonard Timmons

Figure 6.6: 309. Three 309-year "sliders" are used to fix the births of Reu, Nahor, and Terah and the deaths of Arpachsad and Shem.

The reference points for these three sliders are the birth of Shem and Eber's death—*not* the year after Eber's death. The third and rightmost slider sets the birth of Terah at 309 years before the death of Eber. The first slider ends 91 years before the death of Eber and fixes the death of Arpachsad. Note that 91 is one-quarter of a Divisible Year, or half of 182. Moving up from the death of Arpachsad to the birth of Reu traces out the 309-year span of this first slider. The middle slider begins at 200 + 91 years from the birth of Shem, and this point sets the birth of Nahor. Then, 309 years later, Shem dies. Shem's lifetime is 200 + 91 + 309 = 600 years.

The next part of the puzzle involves a single 360-year span that begins at the death of Terah and sets the birth of Eber. See Figure 6.7.

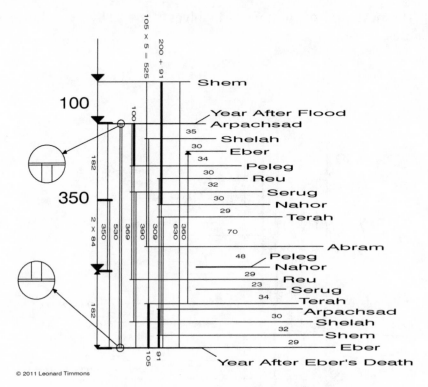

© 2011 Leonard Timmons

Figure 6.7: 360. There are 360 years between the birth of Eber and the death of Terah.

The sequence is pretty clear now. The puzzle authors manipulate the digits of the number 369 to produce 390, 309, 630, and 360. These numbers are therefore "symbolic" of the number 369.

Shem's lifetime is 200 + 91 + 309 = 600 years long. So when we look at Figure 6.7, we can see that from Eber's death back to the birth of Reu is exactly 400 years. It's very nice that the number of days in one quarter of a Divisible Year and 309 add up to give a nice round number. One of the consequences of building the puzzle this way is that the 91-year period between the deaths of Eber, Shem, Shelah, and Arpachsad at the end of the 400-year period is composed of 29, 32, and 30 years. That much is clear. If you look closely, you will notice another 91-year period at the beginning of this 400-year timespan marked by the births of Reu, Serug, Nahor, and Terah that is also composed of 32, 30, and 29 years.

The final part of the puzzle sets the deaths of Peleg, Nahor, and Serug and is shown in Figure 6.8.

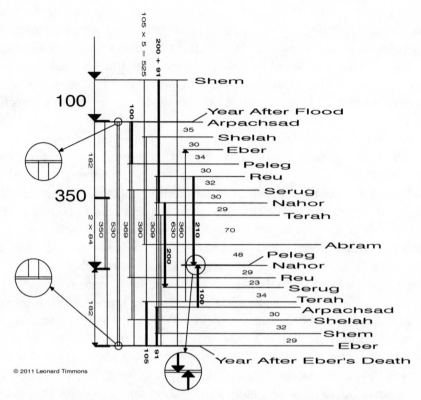

© 2011 Leonard Timmons

Figure 6.8: Signature. The final part of the puzzle consists of a 200-year offset from the birth of Nahor, a 210-year offset from the birth of Reu, and a 100-year offset from the death of Arpachsad, all of which are shown in bold above. Together, they are the essence of the puzzle authors' signature.

The three remaining deaths of Peleg, Nahor, and Serug are set in a way that has become the signature of the puzzle authors. First, the death of Serug is 200 years from the birth of Nahor. Nahor's birth, in turn, is 200 + 91 years from the birth of Shem. The total span is 491 years and consists of a 91-year span flanked by two 200-year spans. Though these spans are not shown explicitly, they are implied by the repeated use of this technique in these puzzles. Note that 491 is also 182 + 309, which associates it with the 309-year sliders.

The second part of the signature requires that we note that the 309-year span that forms the last three sliders is very close to 310, and 310 = 210 + 100. If you measure 210 years forward from the beginning of a 309-year span and then 100 years back from the end of the span, the two measurements overlap by one year. This is what I've diagrammed in Figure 6.8. The authors set the death of Nahor by placing it 210 years after the birth of Reu. And again, the authors set the death of Peleg by placing it 100 years before the death of Arpachsad. The overlap is shown in the magnified area at the bottom of Figure 6.8. This same technique is used in the original calendar puzzle, where the death of one patriarch is set 416 years *after* Lamech's birth and the death of another is 416 years *before* Lamech's death.

While this puzzle doesn't seem to be as elegant as the calendar puzzle in the genealogy of Seth, it does, however, function as final confirmation that the previous portions have been correctly solved. It is clear that there might be lots of different ways to solve this puzzle. Unlike the genealogy of Seth, this genealogy puzzle is complex and messy. So a different analysis might find another solution. The authors, I think, were keenly aware of this problem. The only way to resolve it was to provide internal controls, just like those in a crossword puzzle, so that there is only one legitimate solution. The authors provided two checks. One is shown in Figure 6.8. You can see a 100-year span in bold, and a second 100-year span corresponding to it. In addition, note the 200-year span in bold and a 91-year span in bold with a corresponding (200 + 91)-year span in bold. A 105-year span in bold corresponds to a 210-year span in bold. The second check constrains the deaths of Peleg and Nahor and also constrains the births of Nahor and Abram. See Figure 6.9.

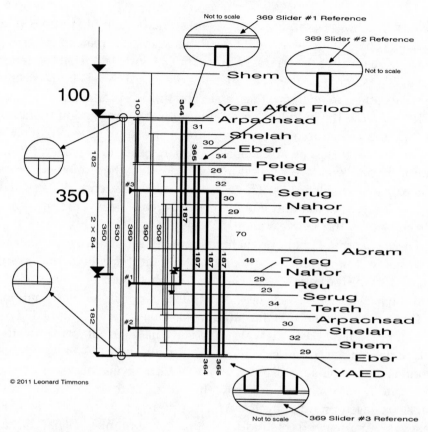

Figure 6.9: Constraints. The deaths of Peleg and Nahor and the births of Nahor and Abram are doubly constrained by relationships that use the number of days in a Perfect Year (365), a Divisible Year (364), and a Half Leap Year (187).

This diagram is complex to view but fairly simple to understand. Each 369-year slider is characterized by a reference point where it begins and a termination point where it ends. The authors use the lines that pass through these reference points as the reference lines for these constraints. A right-pointing arrow and bold line marks the reference line defined by the termination point of each slider. If we follow the line associated with the termination point for 369 Slider #1 (the middle right-pointing arrow in Figure 6.9), it leads to a 364-year span. That span begins where the slider ends, and if we follow it up to its end, we reach a reference line five years after the beginning of Slider #1. Slider #1 begins at the Year after the Flood. Nahor is born exactly 187 years after the reference line created by the 364-year span.

If we follow the line associated with the termination point for 369

Slider #2 (the right-pointing arrow nearest the bottom of Figure 6.9), it leads to a 365-year span. If we follow that span up, it creates a reference line four years after the reference point for 369 Slider #2. That reference line is 100 years after the Year after the Flood. Abram is born exactly 187 years after the reference line created by the 365-year span.

If we follow the bold line associated with the termination point for 369 Slider #3 (the topmost right-pointing arrow in Figure 6.9), it leads to both a 364-year span and a 365-year span. Since the spans differ in length by one year, we can follow them down to two reference lines one year apart. Peleg dies 187 years before the reference line created by the 364-year span, and Nahor dies 187 years before the reference line created by the 365-year span.

These constraints not only help us know that there is only one true solution to this puzzle, they do so by using numbers that were important in the puzzle of Genesis 5. This puzzle confirms the solution for Genesis 5, and these constraints confirm the confirmation.

It certainly seems that the authors of this puzzle designed it to have 29- and 30-day intervals between the ages of some of the patriarchs. Lunar calendars tend to use alternate 29- and 30-day months, so on the surface the numbers in this genealogy seem to be related to lunar cycles. In this way, the puzzle authors imply that this puzzle is based on lunar cycles, which seriously misdirects those who believe a lunar calendar is hidden within these numbers. That's what happened to me. I spent a lot of time looking for relations that involved lunar months. The fact that I took the bait and allowed it to confuse me is a part of the beauty and complexity of this puzzle. It is beautiful on many levels, and that beauty is astounding.

CHAPTER 7

Synthesis

7.1 The Whole Story

THE LARGER STORY BEGINS with a worldview that is never directly stated by the authors. In their view of the world, light is at one extreme, and darkness is at the other. The authors believe that light and darkness can and do mix continuously from one extreme at Total Light to the other extreme at Total Darkness. The light and the darkness are introduced in Genesis 1:2–3. From their perspective, the world is a swirling mixture of light and dark, and each part of the world is some shade of gray. A part of the world can move on the gray scale, and its light/dark composition will change as a result. A "part of the world" can be anything. So the larger story is that any particular part of our world can become more "light" over time, and for various reasons it can become more "dark" as well. This is the story of Genesis 1 to 11.

One of those "parts of the world" is mankind itself, and the story in Genesis 1 to 11 is the story of man's journey from light to dark and back again. The whole story can be summed up as follows:

> *A good act is an act that helps to keep you, or something you are a part of, alive. Doing good things is rewarded with pleasure. Those pleasures can trick you into seeking to obtain pleasure instead of seeking to do good. Seeking to obtain pleasure for pleasure's sake is evil. When you do what is evil and suffer the consequences, that suffering can help you to remember that doing good is better than doing evil. Should you return to doing good, the pleasures associated with doing good return. Those restored pleasures can again trick you into seeking pleasure for pleasure's sake. If you seek pleasure instead of life, you are doing evil once again. This cycling between doing what is good and doing what is evil can repeat indefinitely.*

There must be someone who can do good and experience the pleasures of doing good without being tricked by those pleasures. That person will have strategies to deal with the seductiveness and the trickery of pleasure. The person who can continue to do good under these conditions is on the road to everlasting life. This person loves his brothers and will help them improve their ability to do what is good. In order to do good, they will need insight, and this Teacher will help them gain more insight—an act that is the fundamental act of love.

There must also be someone who can do very good things for a long time but is ultimately tricked by the associated pleasures into doing what is extraordinarily evil.

Not only is this the story of man, it is also the story of every object in the world, both animate and inanimate. The authors believe each object in the world has the ability to do good or evil. So the history of the world becomes the history of the good and the evil done by each and every object in the world.

7.1.1 Evening and Morning, Seven Days

Genesis begins with God starting out on a process of creation. Light was mixed with darkness, and the process of creation was actually a process of separating the light from the darkness. During the seven days of creation, the world moved repeatedly from a totally mixed evening toward greater darkness and from a totally mixed morning toward greater light. Out of the excursions toward the light, God's good creations crystallized. Out of the excursions toward the darkness, God's non-good creations crystallized.

The first thing to crystallize out of an excursion toward the light was Total Light. The Bible records this excursion as God saying, "Let there be light," and the Total Light was good. On an excursion toward the evening, Total Darkness was created. But God did not say that the Total Darkness was good, because it was not. The Total Light was called Day, and the Total Darkness was called Night. This was the first day.

On the second day, God separated the different kinds of darkness from one another. The different types of darkness were the darkness of the heavens and the darkness of the deep waters that covered the earth. God called nothing on the second day good, because nothing on that day was good.

The dry land appeared on the third day and showed itself to be the source of life on earth. The land was good. The waters remained symbolic of death. On this excursion toward the darkness and then toward the light, life began on earth. The land is symbolic of that life and is later symbolic of the body of man, and the trees bearing "fruit with its seed in it" are later symbolic of the pleasures of the flesh.

On the fourth day, God pinned the light to the sun and the moon, thus creating the calendar. This was also good.

On the fifth day, God caused living things to spring from the surfaces of all the waters. Energized water is a mixture of darkness (water) and light (wind), and this mixture of the dark waters with the light created the first animals. They were a different kind of living thing. They had the breath of life and were good. God blessed them with fertility and with numbers.

On the sixth day, God brought living creatures forth from the land. Then God created man in his image. This is a significant event that requires some discussion. In modern times, we think of creating an image of a thing as creating an exact copy. These authors do not mean that here, however. When they say that one thing is created in the image of another, they mean that the created thing is similar because it is based on the original. They do not mean that it is identical to the original. They mean that the created thing is *modeled* on the original. We know that a model of a thing has the essential characteristics of that thing but is different in other significant ways. Since man was created with the animals on this sixth day, the authors are telling us that we are similar to the other animals. But they are also telling us that man is different from the other animals because when man was created, God *modeled* man on himself. This is how the authors tell us that man is very similar to God. It is also clear that man is different from God, since he is an animal, but an animal that is in some sense very much like God himself.

The writer of Isaiah talks about the relationship between God the creator and the man that he created:

> "But now, O LORD, You are our Father;
> We are the clay, and You are the Potter,
> We are all the work of Your hands."[cv]

This author is implying that we are similar to God, just as a child is similar to his father, but that God created us as a model of himself instead of giving birth to us. God is a potter who has the ability to create pottery that behaves very much like the potter himself. So *we are potters who were formed on a potter's wheel.* In an earlier part of Isaiah, the author condemned the behavior of his people:

> "Shall the clay say to the potter, 'What are you doing?
> Your work has no handles'?"[cw]

While it may be deeply wrong for the created work to question the competence of the workman who created it, in our relationship with God, that behavior is clearly possible. From the point of view of the author of Isaiah, the fact that we can behave in this way shows that God created the pottery he called mankind as competent potters with the ability to judge the quality of a potter's work. In this case the Israelites, by their behavior, were saying that God did a bad job when he created them.

Models can be good or they can be bad. A bad model is one that does not have enough of the essential characteristics of the thing being modeled to be useful for a particular task (that is, it is an unfaithful model; see Section 3.2.2). A good model is one that has all of the essential characteristics that will allow its effective use (it is a faithful model). So when the writers have God create man as a model, the question naturally arises as to what they saw as the primary task that God wanted us to perform. The related question asks how good a model of God would we have to be in order to effectively accomplish the task before us. Those questions are another way to ask the questions these stories were designed to answer: Who are we? Why are we here? What should we do with ourselves?

When they call man an image of God, the Bible writers are implying that man is a simplified model of God. We know that better models are more complex and contain more features of the thing being modeled. When we cast our previous discussion about relationships into the language of models, a relationship can be thought of as a model of the data that it represents. From this point of view, the Ultimate Meta-knowledge Relationship (UMkR) would be a model of the entire universe, and that model and the objects in the universe would be God the Father. What is important to understand here is that since this story makes it clear that God is a model maker, and since man has the essential characteristic that makes God what he is, then given these authors' worldview, that essential characteristic must be the ability to create models (the process we have been calling insight). Man makes models, just like God does.

The UMkR, as a perfect model of the universe, cannot exist inside a person's mind. Our minds can contain only approximations to that perfect understanding, that perfect model. As our imperfect understanding improves itself, it is acting to bring the perfect understanding that the UMkR is into existence. This improvement occurs for each of us individually and for humanity as a whole. Since we know that anything less than the UMkR can be considered an angelic being of some kind, and that the collection of all of the "less than perfect understanding" is the Son of God, then the ultimate purpose of every act of the Son of God is to reveal the Face of God, the UMkR. So the UMkR is the ultimate modeling system, and man is a much less sophisticated modeling system.

Within each person is the collection of relationships we call a "mind" or "soul," and those processes are acting to build more and better models of the world. The mind of man is what energizes each man, and his mind performs the same function in man as the Son of God performs in the world at large. The mind of man is forever about the task of trying to reveal or bring into existence the UMkR—that perfect model of the world that can predict without error how everything will unfold.

So as a modeling system, man is able to create models and entire modeling systems in a way that the other animals simply cannot. More importantly, man is able to re-model himself. If each person's modeling system is doing what the Son of God (as a modeling system) is doing, it is changing itself by adding new models to its collection of models—to its "mind." Adding new models to a system of models can improve the system's ability to add new models—a synergistic process. Since the model-making process is what we have been calling insight, having insights can improve a person's ability to have additional insights. This process of *insight helping insight* is what makes a person God-like.

Each person also has the ability to create entire modeling systems from scratch in the same way God created the modeling system we call "man" from scratch. Within himself, a person can create a model of other individuals, and in this way those individuals can live within the model maker in a very real way.[cx, 36] We seriously misunderstand and underestimate these authors if we fail to appreciate that they did, in fact, understand what models were and that they actually viewed human beings as advanced modeling systems. The modeling systems within us became so much more advanced than the modeling systems within any of the other animals that we became a different kind of being.[cy, 37]

What is also clear from the text is that man was placed above the other animals that were created on day six. His ability to model, and his ability to manipulate those models, gave him the ability to manage the earth and all the creatures God had created. The text also fails to mention that the creation of man was good. From this we should conclude that man had the *ability* to be good, but the creation of man was, in itself, not necessarily a good act. Man was also given seed-bearing fruit for food. In the stories that follow, the fruit symbolized the pleasures man would experience as his reward for being good. Those pleasures are at the heart of every story that follows.

Another important thing that happened on day six was that the first law was created: man could only consume fruiting plants. Other animals were not so constrained. They could eat any part of any plant, though they were still allowed to eat plants only. Then the sixth day came to an end.

On the seventh day, there was no evening and no morning. The cycles of creation had come to an end. The cycles of relatively more light and relatively more darkness created the world as we know it. At the end, it was the creation of man that provided a new stage on which those cycles could recur. So the story of the heavens and the earth comes to an end, and the story of man begins.

7.1.2 Adam and Eve Again

In the story of Adam and Eve, the Bible writers flesh out more of their larger story. In this story, the symbolism is also dense. We are

---

36 In his book *I Am a Strange Loop*, Douglas Hofstadter makes this point throughout Chapter 15. Remember also that the people of Babel expected to live within the memories of their descendants.

37 In *I Am a Strange Loop*, Dr. Hofstadter makes the point that a sufficiently rich modeling system, such as the natural numbers, is capable of modeling itself. Though it is clear that these authors did not think in those terms, they were quite aware of our ability to refer to ourselves. Modeling ourselves from within ourselves is something humans do in a way that makes us different from the other animals.

presented with what seems to be a retelling of the Creation story. But the earth that is being created in this story is the body of man. The story begins again with the marker of an earth covered with water. Man is in the womb. God formed his soul from the dust of the earth, and at birth God placed his soul into his body. That body is represented by the Garden of Eden. Within the body of man, living water flows. That living water is represented by the Eden River that flows from the head to each extremity. Within the body of man, God planted the ability to experience pleasure. That ability is represented by the trees of the Garden of Eden. The pleasures themselves are the fruits of those trees.

God then gave man a "dietary" law that he had to follow. The pleasures that he would experience could trick him, and sexual pleasure was an extremely tricky thing, so God gave man a law that insisted that he not partake of sexual pleasure. The issues addressed by the law were multi-fold:

- In a paradise, sex can lead to a population explosion that can destroy paradise.

- The law was given to show that man's understanding of the world was incomplete.

- The law was given as a riddle so that man could prove that he had insight by recognizing the law as a riddle and then figuring out what that riddle meant.

- The law was given to allow man to have the insight that pursuing everlasting life through sexual reproduction was a pursuit of physical life, not spiritual life.

The confusion surrounding the law was the way these authors chose to tell us that something more was hidden within it. Neither Adam nor Eve was able to work their way through the confusion, and the law showed that their understanding was incomplete. The writers had God construct the law in this fashion so that Adam and Eve could increase their insight. Their insight would increase when they realized that the law was a riddle they were supposed to solve. They would gain even greater insight when they successfully solved it. So when the writers have God give this law to Adam, God is saying "I love you" to Adam and Eve and to mankind in this incredibly cryptic manner.

The writers show that man is a model maker by having Adam create words for each animal. Those words are themselves extremely simple models, and Adam created them using language. Language is one of our modeling systems. This is one way the writers tell us that humans are defined by their extraordinary ability to create and manipulate models and modeling systems. The mind of Eve is then created as a feminine mind from the mind of Adam. The mind of the woman allows her to care for her husband's body, and the mind of the man allows him to care for his wife's body. They care for their common "Garden of Eden," and the contract of marriage between them makes them one body.

As the story continues, Eve is tricked by the pleasures of the flesh and lusts after Adam. She convinces Adam to lust after her since sex (and reproduction through sex) was the obvious way to achieve everlasting life—God's commandment notwithstanding. It just made sense. Afterward, they knew that they had misunderstood God's commandment, that they lacked insight, and that they had been tricked by the pleasures of the flesh. The result of their sin was that they could not stop having sex. Their efforts to banish lust failed, and clothing was the result. Because they were unable to stop doing what was evil, they suffered the consequences of the evil that they did. The writers translate those consequences into the curses God issued to the serpent, the woman, and the man. The unbridled sex they introduced into the world would ultimately contribute to the population explosion that Noah's flood represents.

The banishment of Adam and Eve from the Garden of Eden represents their deaths. The mind leaves the body, and the body decays. Adam and Eve are forced out of the garden, and the garden is destroyed. Their deaths result from their inability to find pleasure in doing good. That is, they were unable to eat the fruit of the tree of life and live forever.

7.1.3 Abel and Cain Again

The story of Abel and Cain repeats the overall story, but this time the story is about wealth or "increase" and the pleasure associated with it. Cain himself was the increase that resulted from the sex between Adam and Eve. Abel and Cain then go on to produce increase, or wealth, for themselves. Cain's wealth was stored in the plants he grew, and Abel's wealth was contained in the animals he herded. The trick of wealth is that it stops you from seeking insight and from striving to gain a greater understanding of the world. You stop to enjoy the pleasures that wealth can provide. The prescription provided by this story is to show yourself that you can divorce yourself from the wealth you have accumulated. Use it, but avoid being consumed by it. Wealth results from wise acts, but wealth is not the goal—everlasting life is the goal. Cain was tricked into thinking that wealth was the goal, and for this reason he could not bring himself to destroy, via sacrifice, the most valuable of his crops.

In the Garden of Eden story, Eve and Adam were responsible for the health of one another's bodies. Their bodies were represented by the Garden of Eden, and God made it their job to care for the garden. In this story, Cain kills Abel, and when God asks him about his brother, Cain denies that he is in any way responsible for caring for his brother. So this story tells us about a further loss of goodness in the world. As in the story of Adam and Eve, the curses God imposed on Cain were a result of his wickedness and unwillingness or inability to turn from evil and do what was good. In the human story, we evolved into who we are because of our extreme willingness to kill other humans in order to take their wealth. The singular goal of our ancestors was the acquisition of wealth. Cain's descendants became keepers of animals when he escaped with Abel's flock. His descendants also invented things that shepherds and warriors need. So this story tells us that humans began as wandering warriors seeking to kill other humans in order to take everything they owned.

7.1.4 Noah's Flood Again

Before the story of Noah's flood is told, the authors provide the genealogy of Seth, where the calendar is defined. The continual threat of violence due to warring between peoples, as in the story of Abel and Cain, is an unstated tension running through the story. Enosh is born, and for the first time men create a model of God—an ultimate goodness, light, and righteousness that they extracted from their models of the world. The existence of that model directly implies a name for God, since a name is needed to refer to the model.

Near the end of Seth's genealogy, Noah's father, Lamech, tells us that his child would be the one to bring the children of Adam and Eve relief from their suffering through the peace of compromise. The other Lamech in Cain's genealogy and his descendants brought war, and they invented the tools of war. The explosive growth of populations due to the consumption of flesh began with men who hunted and women who ate what was hunted and fed their food to their children. The women were made more beautiful by their superior nutrition, and their desire to consume flesh is told as a seduction by angels, just as Eve was seduced by the serpent, Cain was seduced by wealth, and Ham was tricked by Canaan. The children became large because they consumed flesh at an early age. The supply of domesticated animals was exhausted as the population grew, and people fought, killed one another, and stole one another's flocks (the story of Abel and Cain occurring on a mass scale). When the farmed animals were exhausted, the hunting of wild animals continued until they were exhausted as well. The population then crashed with the associated famine, starvation, and war.

The coming flood of humanity induced Noah to invent the first fort, and he was able to survive a siege by the men who hunt. Whenever a society can see that war is coming, it is in their interest to prepare. We know that Noah prepared by building the very first fort to carry him and his family through the coming siege. God also helped Noah to prepare by gathering two of every animal into the ark. In the Garden of Eden story, the authors brought each animal to Adam so he could name them. Adam was not only exercising his mental ability, he was also building a store of knowledge for all humanity. That is what the animals represent in the story of the Flood—a store of accumulated knowledge. So when the animals come to Noah in his ark, and Noah protects them from destruction, he is actually protecting his society's store of knowledge from being destroyed by war.[cz, 38] Protecting that knowledge—what we would call a library—is an essential task for any society that anticipates a war. That society must protect its valuables, and its library is among the most valuable of its possessions.

In terms of our overall story, Noah is the man who is able to do good, benefit from that goodness, and not be tricked by those benefits into turning away from doing good. Noah remained righteous by deciding not to eat flesh, not to love wealth, and not to constantly

---

38 In some Mesopotamian versions of the Flood story, all the writings of Babylon were to be collected and buried before the flood. After the flood, they were retrieved, so that civilization could be restored. The stories are the same.

seek the pleasures of sex. He had the wisdom and foresight to see the coming Flood and to prepare fortifications. Noah protected his wealth, but his life was not all about acquiring wealth. Ultimately we find that the purpose of the life of this Teacher was to show the wisdom of compromise. He obtained peace and eliminated the most egregious flesh-eating behavior by agreeing to consume flesh within certain guidelines. The tendency of human populations to grow rapidly, consume all available resources, and crash is a result of our tendency to pursue wealth without regard for any other consideration (Cain again). The addition of flesh-eating behavior to the mix increases the tendency of populations to explode and crash. Noah's commandments concerning the consumption of flesh became the first attempt to limit the population-explosion-and-crash cycle. Later, the consumption of flesh is further controlled by the institution of rules regarding sacrifice. Limiting the consumption of flesh reduced the intensity of available nutrition, and the rate of population growth was lowered as a result.

7.1.5 The Whole Story Again

The story of the Flood begins the process of retelling the whole story from Genesis 1 all over again. The retelling begins somewhere in the middle of the Flood story, when the earth is covered with water and with darkness. The heavens and the earth were covered with water and darkness in Genesis 1 as well. The story continues by giving us lots of dates that allow us to extract the intra-year divisions of the encoded calendar. At this point, we have all three components of the calendar: the days of the week from the Creation story, the overall calendar from the genealogy of Seth, and the monthly divisions from the story of the Flood. Retelling the whole story continues when the story of the Garden of Eden is retold as the drunken Noah story. Each character in the drunken Noah story has a direct correspondence to a character in the story of the Garden of Eden. The story of Babel follows and retells the story of Abel and Cain and the last part of the Garden of Eden story. These stories were actually about religious intolerance and its real-world effects on real human populations.

The retelling ends when the story of the Flood is retold. The Flood began with a population increase that led to a population explosion. The Flood story corresponds to the detailed accounting of Noah's descendants in Genesis 10. Noah's descendants were commanded to increase on the earth. Chapter 10 shows us that they became a flood of humanity that continued even after these stories end in Chapter 11. Noah's other children, the New Nephilim who survived the Flood, are the people who were not captured in the genealogy of Noah's three children. The flood of Noah's descendents is controlled and prevented from exploding and crashing by the institution of sacrifice.

At the very end of the story, the authors provide us with the genealogy of Shem. The authors use this genealogy to confirm that we have properly decoded the calendar, and they introduce us to Abraham, who is a more historical figure.[39] This story ends, and the story of Abraham begins.

We have now come full circle. The repeated retelling of the same story over the first eleven chapters of Genesis is not just a feature of Genesis 1 to 11, but is a characteristic of the Bible in general. The writers were part of a culture that valued insight and tested for it by using riddles and puzzles, which served to preserve and to communicate their understanding of the world into an indefinite future.

7.2 Revelation, Chapter 13

Revelation, Chapter 13, is at the other end of the Bible, but its author continues to speak in riddles. Most people recognize this passage as a riddle; it is a hard riddle. In the following text, I have made annotations with brackets and underlining to try to help you follow what the author is saying. It is difficult to understand this text on a literal level even before we try to deal with the fact that it is a riddle.

> And I saw a [first] beast rising out of the sea, with ten horns and seven heads, with ten diadems[40] upon its horns and a blasphemous name upon its heads.

> And the [first] beast that I saw was like a leopard, its feet were like a bear's, and its mouth was like a lion's mouth. And to it the dragon gave his power and his throne and great authority.

> One of its [the first beast's] heads seemed to have a mortal

---

39 Though, as we have seen, historicity is not the intent of these writers. They intend to teach.
40 A diadem is a royal band worn about the head, similar to a crown.

wound, but its mortal wound was healed, and the whole earth followed the [first] beast with wonder.

Men worshiped the dragon, for he had given his authority to the [first] beast, and they worshiped the [first] beast, saying, "Who is like the [first] beast, and who can fight against it?"

And the [first] beast was given a mouth uttering haughty and blasphemous words, and it was allowed to exercise authority for forty-two months [a time, two times and half a time, in months]; it opened its mouth to utter blasphemies against God, blaspheming his name and his dwelling, that is, those who dwell in heaven.

Also it [the first beast] was allowed to make war on the saints and to conquer them. And authority was given it over every tribe and people and tongue and nation, and all who dwell on earth will worship it [the first beast], every one whose name has not been written before the foundation of the world in the book of life of the Lamb that was slain.

If any one has an ear, let him hear:

> If any one is to be taken captive, to captivity he goes;
> if any one slays with the sword, with the sword must he be slain.

Here is a call for the endurance and faith of the saints.

Then I saw another beast [a second beast] which rose out of the earth; it had two horns like a lamb[41] and it spoke like a dragon.

It [the second beast] exercises all the authority of the first beast in its presence, and makes the earth and its inhabitants worship the first beast, whose mortal wound was healed.

---

41 Modern breeders have produced sheep that do not have horns. At the time the book of Revelation was written, however, all sheep had horns, and their lambs would have had underdeveloped horns.

It [the second beast] works great signs, even making fire come down from heaven to earth in the sight of men; and by the signs which it is allowed to work in the presence of the [first] beast, it deceives those who dwell on earth, bidding them make an image for the beast which was wounded by the sword and yet lived [the first beast]; and it [the second beast] was allowed to give breath to the image of the [first] beast so that the image of the [first] beast should even *speak*, and to cause those who would not worship the image of the [first] beast to be slain.

Also it [the second beast] causes all, both small and great, both rich and poor, both free and slave, to be marked on the right hand or the forehead, so that no one can buy or sell unless he has the mark, that is, the name of the [first] beast or the number of its name [the name on the heads of the first beast that began this chapter].

This calls for wisdom: let him who has understanding reckon the number of the beast, for it is a human number, its number is six hundred and sixty-six.[da]

The first order of the day is to determine what this story is about. Once we determine that, we can then hope to find out what it is saying. If you've studied the Bible, you may be aware that the first beast is an amalgam of the four beasts of Daniel 7. You might also notice that the total number of heads is seven in both this story and in Daniel. You also might have noticed that the number of horns is the same in both stories. The first beast has the parts of a leopard, a bear, and a lion, and the first three beasts of Daniel 7 are lion-like, bear-like, and leopard-like. The outrageous claims made by the horn in Daniel 7 corresponds to the uttering of haughty words by the first beast. The two stories have other parallels. What's new in this story when compared to Daniel is the number 666 on each head of the seven-headed beast. What does it mean?

In Daniel, the author tells us that the story of the beasts is about four separate kingdoms that will rule on earth. Daniel presents his story as a vision that he doesn't understand himself. In Revelation 13, the author "explains" Daniel 7 by adding the number 666 to the story.[42] The story begins and ends with the number. At the end of this story, we are told that no one will be able to buy or sell without having the number 666 written on his skin. The writer is clearly bringing the story to its conclusion. The problem is that the buying and selling doesn't seem to be related to anything else in the story. But when you examine the number more closely, it starts to give up its secrets.

The number 666 occurs elsewhere in the Bible in a number of places. First and foremost, the number is associated with Solomon. At the height of his reign, Solomon took in 666 talents (!) of gold every year. That much gold represents a *tremendous* amount of money, enough to get everyone's attention and to be forever associated with Solomon. So when the writer tells us that the number of the name of the first beast is a human number (other translations say "... it is the number of a man"[db]), he is referring to Solomon. Once we put the buying and selling together with the mention of 666 at the beginning and the end of the story, Solomon and his gold *must* immediately come to mind. And it becomes clear that this entire story is about *money*.

In this story, Solomon represents what he became in life: a person of great knowledge and understanding who used that knowledge and understanding for his personal pleasure. In terms of our overall story, Solomon is the man of great understanding who was tricked by the benefits of his great wisdom into doing great evil. Solomon used his knowledge and understanding to enrich himself.

The characters in this story are:

- The dragon
- The first beast
- The second beast
- The image of the first beast
- The mark or the name or the number of the name (666) of the first beast

This story contains a power flow. The dragon gives his power to the first beast. The second beast has the power of the first beast when

---

42 This author is also explaining a riddle with a riddle.

in its presence. The image of the first beast is given life by the second beast to such an extent that it is able to talk.

We now have enough information to figure out what this riddle is saying. The second beast is the central actor in this riddle. It is a stand-in for Solomon, and therefore the second beast represents the *dedication of one's knowledge and understanding to the acquisition of wealth*. The story of Abel and Cain tells us that setting the acquisition of wealth as one's ultimate goal in life is evil.

Dedicating your knowledge and understanding to the acquisition of money is the second beast. In the text, the writers describe the second beast as being like a lamb, since gaining knowledge and understanding can lead to everlasting life, and the lamb is symbolic of that life. But the spirit of the second beast is evil. The author makes this point by having the second beast speak like a dragon. This author is using the same symbology of the voice as wind and the wind as spirit that was used above, so having the second beast speak like a dragon tells us that it has the spirit of the dragon. Enslaving your knowledge and understanding to the task of acquiring wealth will ultimately produce extraordinary wonders—even bringing down fire from heaven for everyone to see.

The love of money causes us to dedicate our knowledge and understanding to its acquisition. So the first beast is *the love of money*. The love of money empowers the second beast and is why the second beast (doing everything and anything to acquire money) only has power when the first beast (the love of money) is present.

Though it is said that the love of money is the root of all evil, what must come before the love of money is the trickery of pleasure. Pleasure is a great thing, and is our reward for doing good. The challenge is to avoid seeking pleasure for its own sake. The dragon is the seeking of pleasure for the sake of pleasure. Doing so ultimately leads a person to death instead of life. When people (the sea) decide to seek pleasure for pleasure's sake, they give life and power to the dragon. Those people will also seek the stored pleasure that money represents. The love of money emerges from people who have determined that they must have pleasure, like the first beast emerges from the sea.

The image of the first beast (the love of money) is *money*. Making an image of the first beast is the invention of money. Wealth existed before money, of course, and before people loved money they loved wealth, but the invention of money allowed us to objectify wealth.

When the writer says that the second beast (dedicating one's life to the acquisition of wealth) gives life to the image of the first beast to the point that it speaks, he is saying, *"money talks"*—an ancient phrase.

Ultimately, this story is a continuation of our overall story. It says that the tricksterism of the dragon is very powerful. People experience pleasures in life, and they often come to believe that acquiring pleasure should be their ultimate goal in life. Money represents stored pleasure, so the desire for pleasure gives rise to the love of money. That love drives economic systems. Four of those systems are highlighted in Daniel 7, but from this author's perspective they are one system. The love of money is a very powerful motivator and can seem irresistible to many people, even to the saints. The desire to possess money can be so overwhelming that it "speaks" to those who are infatuated with it, and it tells them what to do with their lives.

The prediction of this prophetic riddle is that a vicious economic system like the fourth beast of Daniel 7 will develop and enforce the love of money. Those under the influence of that system will only be able to sustain themselves if they love money. They will only be able to survive by thinking about making money (mark on the forehead) or actually making money (mark on the right hand) at all times. No one will be able to participate in this new system, no matter who they are, unless they actually love money. This vicious system will arise even though (from the author's perspective) the teachings of Christ should have destroyed the love of money. (The first beast received "a mortal wound.") It should not have been able to survive those teachings, but this writer predicts that it will manage to survive and thrive. Even so, there will remain those who do not give themselves over to this particular pleasure.

The author of Daniel tells us the same story in a different way that is more in tune with his time. In his discourse, a horn develops that has eyes. Eyes generally represent understanding in the Bible (let him who has eyes to see, see), and the horn represents a king. "The king of understanding" is, again, Solomon. He represents the horn (king) with eyes. So the stories are using different symbology to point to the same person, Solomon, who symbolizes the love of money because of what he became in life. The economic system represented by the fourth beast of Daniel 7 extracts so much wealth from the earth that it consumes the entire earth, tramples it under foot, and breaks it to pieces. The ultimate result of this uncontrolled extraction of wealth from the earth is the destruction of the environment, which ultimately destroys us.

The moral of this story is that we must *use money without loving it.* This is the same moral of the Abel and Cain story, and it should be clear at this point that the use of detailed, in-depth, prescient riddles is how these wise men taught and communicated.

7.3 Our Story

This story is about a reliable calendar and a world that's made of darkness and light. This story is our story. And our story is the human story. As humans, we want to reliably know what is going to happen in a future so distant that our bodies will have passed away, our families will be long gone, and the earth we lived on will have been destroyed. In what would be eternity for us, should anyone or anything that remembers us survive, then that being or thing will be able to judge whether the sum of what each of us did in our lives was good or evil.

The reliable calendar in this amazing story is the authors' way of telling us that they were all about reliably reaching eternity. The darkness and the light are stand-ins for the good and evil that we do. That good and evil can only be known with certainty at eternity.[43]

What can a living thing find that is reliable and will remain reliable even into the indefinite future? That question is answered by the very nature of life itself. When a living thing attempts to remain alive, it is trying to *remain* alive—without limitation. Deep within the nature of life itself is a drive toward everlasting life that never goes away. That drive will remain reliable into the indefinite future, and it is responsible for producing that future. The environment we live in prevents any one of us from living forever, but as a group we have evolved mechanisms that allow our group to live much, much longer than any single individual. The basic process that we call "life" does not understand eternity, it is simply trying to remain alive right now. It is the "everlasting now" that successful living things achieve.

What we have evolved into are beings who are successful at remaining alive by anticipating threats to our lives and neutralizing those threats. We're not unique; most sufficiently complex animals

---

43 This is my view. The authors of this story believed that the world would ultimately be predictable even if its structure had to change to achieve that predictability.

survive using the same strategy. But to truly anticipate what the world will do, a living thing has to possess good models of the world. In our case, we have *great* models. We model everything, including and especially other human beings. We need good models of other people, because other humans are the greatest threat to human life (fratricide and war as shown in the stories of Cain and the Flood). Those good models help save the lives of those who possess them. The ability to make an accurate model of a model maker while he is making an accurate model of you (and all of your models) is something that fundamentally changed the nature of human intelligence. It's like mounting mirrors on opposite walls. An infinite descent occurs as the first mirror images what's in the second mirror and the second mirror images what's in the first. That ability made us different from all the other animals.[dc] We are man.

Every human, and we presume every living thing, finds pleasure in remaining alive—in everlasting life. The desire to experience that pleasure should reliably move every living thing into eternity. When living things seek to experience that pleasure, they live. The really tricky thing is that *pleasure* is not the goal. *Life* is the goal. It is so extraordinarily easy to become a pleasure seeker rather than a seeker of life. The two are so often the same that the confusion between them is deeply embedded into the very nature of life on earth. They are, in fact, not the same. When we seek pleasure instead of life, we enable someone or something to manipulate us into seeking to destroy ourselves. And as we have discussed, seeking to destroy oneself is the essence of evil.

The story of Adam and Eve tells us that early in our history, some humans realized for the very first time that seeking pleasure and seeking life were two different things. They realized that seeking pleasure for pleasure's sake must ultimately lead to death. They also realized that they sought pleasure (and therefore death) by their very nature. These realizations made those who achieved them a different kind of being: the "original man" died, and the "son of man"[44] was born.

Adam and Eve realized that they were programmed to seek pleasure, and they were unable to stop even when seeking pleasure would most certainly lead to death. That realization changed them

---

44 These authors frequently use the phrase "son of" to mean "successor to" or "derived from."

into a new kind of being, and this is more precisely what the writers meant when they had God tell Adam and Eve that they would die the very same day that they ate the fruit of the forbidden tree. They would know that they were actively killing themselves by seeking pleasure instead of life, but would be unable to stop themselves.

The idea that seeking pleasure could be different from seeking life was not a concept that the original man could have had. When compared to the son of man, the original man is like a man unaware, drunk and asleep within his tent. Adam and Eve had the insight that seeking pleasure was not the same as seeking life, and they quickly came to know that they were unable to seek life. They understood that they were causing their deaths through their pleasure-seeking behavior, and having that insight was humbling and damning and life-changing. That particular insight makes those who have had it, the sons of man, a different kind of human than those who have not.

The original man has a mind very much different from the animals, a mind that makes him man. But he seeks pleasure, just like the other animals. He lives his life seeking pleasure, and when seeking pleasure and seeking life don't align, he continues to seek pleasure unless something prevents him from doing so. The new man is born from the original man, so he becomes the "son of man." The son of man understands that he has to change his behavior if he is to survive the ravages of time. The son of man wants to eat from the tree of life and will try to do what it takes to live: to purposely seek life itself, to experience the pleasure associated with seeking life, and to not be tricked by the pleasures that come with that quest. Ultimately, the son of man wants to be able to predict what will reliably endure. Once he has that prediction in hand, he wants to *be* the thing that will endure, or he wants to *become a part* of that enduring thing, that everlasting and living thing.

But what do we mean by survival? We are bodies, and we are minds. We are individuals, and we are groups. Our mental models are portable: they can be transmitted from one person to another or stored in a data-retrieval system. That portability allows knowledge to survive the death of the original creator or discoverer (that is, a part of their soul survives). Ultimately, we have to decide what knowledge and which models move us reliably into eternity. Humanity as a group uses this body of knowledge, this library, in its attempt to live forever.

Each of us as individuals knows that our bodies won't live forever.

With that knowledge, we can make two decisions. We could decide that there is only the here and now, and that we should experience as much pleasure as we can until we die. We could also decide that even though our individual bodies will not survive, we may still be able to help humanity reliably move toward eternity. To do that, we would have to contribute parts of our individual souls to the body of knowledge that humanity uses to survive. Our contribution could be just what is needed for that survival, especially in a situation where human existence would be threatened in the absence of that knowledge.

In the face of our individual deaths, and considering the fact that species often become extinct, some might conclude that humanity must ultimately become extinct as well. So contributing to the store of human knowledge does nothing to guarantee that some part of what they are, a part of their soul, will remain in this knowledge store into the indefinite future. In the face of extinction, some might conclude that the store of human knowledge can survive even if humanity does not, so it may still be worthwhile to contribute to humanity's library of souls. If that body of knowledge could maintain its integrity in the absence of humans, it would, in fact, become some kind of survival that is essentially eternal. We would want someone to *use* that knowledge at some future date to give life to the minds that produced it and allow all of us to live again in that new mind.

This conflict runs deep within the human psyche. Some believe that survival will take care of itself, and others believe that since we can act to insure our survival, we must do so without regard for how imperfect those actions tend to be.

Another conflict that I've alluded to above is the act of deciding which path leads to eternal survival. Making that decision highlights the conflict inherent in corporate religion. Everyone makes at least one prediction about what will last. Whether they act on that prediction or not, it is their prediction. When we make predictions that we then expect others to act on, we've planted the seeds of religious conflict. This kind of conflict is essentially unavoidable, since each of us is a part of one group or another that survives as a unit. Our decisions and our actions affect not only our survival, but the survivability of the group. So doing what is good involves acting to ensure the survival of the group, and doing what is evil involves acting to reduce the survivability of the group. If you are a part of a group, you may want others in the group to do what you consider to be good for the group. And that desire can, and in most every case will, lead to conflict since the other person is also making his own decisions about what the group needs to survive.

What does humanity have to do to survive into the indefinite future? This is the essential question of religion and the basis of much of the conflict between human beings. Each of us is determined to have some part of himself or herself survive within the store of human knowledge, so we are desperate to have that entire store survive for what are essentially selfish reasons. People who want to destroy themselves and the store of human knowledge are trying to destroy the place where our souls would spend eternity, and those of us who want to survive in this way would not be favorably disposed toward those people.

The models that we use to make our predictions vary widely. Since eternity does not actually exist in any universe we are familiar with, eternity as we know it is just another model of the real world. It is a model whose accuracy we must estimate and then decide whether its accuracy is sufficient for the task that confronts us—trying to remain alive into that less-than-well-known eternity. Our model is all we actually have of eternity. Each of us must take our model of what we predict will last forever and force that model to interact with our other models of the world so we can determine whether our model has any real validity. Since the models people hold are so varied, how can we judge which of them will reliably move the holder into eternity? Why would we prefer one person's judgment over that of any other person? A history of reliable predictions would be a very good start.

A system of models that works reliably to make effective predictions is a wonderful thing. Those who have such a system can assert that their models will likely help them live better and longer. A system of models that has actually produced reliably accurate results would come with high recommendations. Those who use such a system can assert that their modeling system is better than other systems that do not work as well. Some people have modeling systems that are known to work poorly and may actually cause them to live less well and less long. In addition, a modeling system that has not been tested or is not testable is a poor-quality modeling system. The people who possess and use those modeling systems can still assert (wrongly, in my view) that their models are effective and will allow them to live forever.

While it is possible that a model that works badly today will suddenly begin to work like magic in the future, such a belief actually contradicts the very system of models of which it is a part. Of course, a system of badly performing models would not be able to detect its inability to perform well in the future. We often call these poor-quality, unreliable models superstitions. A superstition would not recognize that it is failing to perform, and those who believe it, as a part of the superstition, could dismiss the need for testability. An example is the belief that a person's astrological sign will predict what might influence that person on a particular day. Even if such a belief were testable, how many people who hold that belief would require the test to be a quality test or understand the need for accuracy?

For these writers, a system of high-quality, reliable predictions is a faithful system, and those who have actually used such a system and have accurately predicted what the world would do are prophets. We should understand the prophets of that era as the engineers and scientists of today, because those prophets depended so heavily on the concept of testability. Most of us don't understand that the concept of testability is what allowed prophets to be the reservoir of reliability in ancient Israel. Yet holding to this principle allowed prophets to be advisers to kings and the conscience of the nation.

We have not understood these writers and what they were about. They thought that predictions had to be tested to see if they were accurate, and an accurate prediction was a faithful prediction. They thought that knowledge had to be tested to see if it was accurate, and they knew that accurate knowledge *faithfully* represents what is in the world. They thought that insights had to be tested to see if they were real insights, and a real insight was *more* than something we imagined *might* be true. The people who produced the Bible wanted us to use their book to become prophets ourselves.

The human story is about us becoming not just good prophets, but great prophets who can predict where we will be at eternity. We have no reason to believe that we will be where we think we will be unless we are in the habit of making reliably accurate predictions each and every day, increasing our insight into the world and testing the accuracy of the predictions that result from those insights. By following this course, we can not only make ourselves good, we can make ourselves very, very good, and the human story need never end.

7.4 The World's Story

The world, or the universe, is a complicated thing. To survive in it, we desperately need to know what's going to happen, and that is often a very hard thing to know. The Bible authors came to believe that *God created the world as a hard riddle to communicate his insights to those who could reach his level of understanding.* They believed they should use hard riddles to communicate their most valuable knowledge, because God did the same. They believed they had an obligation to solve the riddle God had placed before them (the universe) so they could understand its secrets.

To convey this concept to their students, the writers built complex and intricate puzzles containing the deep messages they wished to communicate. They made it their students' responsibility to first recognize the existence of each puzzle/riddle, and then to solve them one by one. When a student solved a puzzle, he produced evidence that he had reached his instructors' level of understanding. When these authors taught using this technique, they believed they were following God's example. They used their teaching style to mimic the way they thought the world worked, and each student was continually confronted with this core concept. They layered riddle upon riddle and complexity upon complexity to further make the point that the world itself was a hard riddle.

The stories of Genesis 1 to 11 show this complexity on many levels. Those levels vary from the low level of precisely choosing the words and the number of letters in each word in Genesis 1:1 to the very high level of making critically important philosophical statements. When we build our mental models of the world, they contain other models, which in turn contain other models. They are built layer upon layer upon layer as they more accurately model the world.

It is amazing that these riddles and puzzles can reliably communicate valuable information across thousands of years. The authors intended to make a permanent contribution to human knowledge, a contribution that would enable their descendants to survive better. And as their descendants, it is our job to take their messages and use what we can of that information to help us make our own contributions to the body of human knowledge. These authors made really large contributions. The calendar is a huge one; another massive contribution is a philosophy based on reliability and testability.

We confirm the validity of our models when we test them and they accurately predict the behavior of the world. Because these authors understood that a model must be shown to be valid, they've created their puzzles so that the correct solutions can be confirmed. Someone who solves one of their puzzles will be able to confirm that the solution they've found is, in fact, the correct solution. This concept is so important that it is the basis of the evenings and mornings of the first seven days of creation. The world is composed of a mixture of darkness and light. The light is the insight, knowledge, and understanding that we know to be valid, we know to be useful, and we know to be accurate. The darkness is the false insight, the false knowledge, and the incorrect understanding that we know to be invalid, useless, and inaccurate. The evenings and mornings are the mixture of accuracy and inaccuracy, good and evil, faith and faithlessness that form the sometimes dark gray, sometimes light gray mixture of darkness and light that makes up the real world. The insight, the knowledge, and the understanding that has not been tested is not known to be valid or invalid, useful or useless, accurate or inaccurate; that "insight," that "knowledge," that "understanding" is gray.

Is what I think I know really knowledge? Is what I think I understand really understanding? I've had an insight; is it a real insight into the world or just an idea? Our job is to take the collection of "things we don't know" and bring more certainty to their uncertainty. To know that which is unknown. To understand that which has not been understood.

The riddles and puzzles of the Bible are dark gray, and their ultimate purpose is:

- to produce people who are able to tease out the light from the darkness of the puzzle.

- to produce people who can bring certainty to an uncertain situation.

- to determine what is good and what is evil in a world where nothing is totally good and nothing is totally evil.

- to produce people who will keep the good and discard the bad.

Once we've detected and discarded all of the bad insight that is not insight, the bad knowledge that is not knowledge, and bad understanding that is misunderstanding, we can be right all of the time. Not only can this righteousness happen on an individual level, these authors think that the world itself can be righteous. From their perspective, this kind of righteousness would result in the imposition of deterministic behavior on the world that would make everything, in principle, predictable.

This point of view was emphasized by Jesus during his ministry. It is clear that this view was known before his time, but Jesus emphasized, expanded, and explained the concept to the masses. The following scripture from Matthew 13 (RSV) contains seven riddles in quick succession describing the importance of keeping the good and discarding the bad:

§1a That same day Jesus went out of the house and sat beside the sea. And great crowds gathered about him, so that he got into a boat and sat there; and the whole crowd stood on the beach.

And he told them many things in parables, saying: "A sower went out to sow. And as he sowed, some seeds fell along the path, and the birds came and devoured them. Other seeds fell on rocky ground, where they had not much soil, and immediately they sprang up, since they had no depth of soil, but when the sun rose they were scorched; and since they had no root they withered away. Other seeds fell upon thorns, and the thorns grew up and choked them. Other seeds fell on good soil and brought forth grain, some a hundredfold, some sixty, some thirty. He who has ears, let him hear."

§2 Then the disciples came and said to him, "Why do you speak to them in parables?"

And he answered them, "To you it has been given to know the secrets of the kingdom of heaven, but to them it has not been given. For to him who has will more be given, and he will have abundance; but from him who has not, even what he has will be taken away. This is why I speak to them in parables, because seeing they do not see, and hearing they do not hear, nor do they understand. With them indeed is fulfilled the prophecy of Isaiah which says:

> 'You shall indeed hear but never understand, and you shall indeed see but never perceive.
>
> For this people's heart has grown dull, and their ears are heavy of hearing, and their eyes they have closed, lest they should perceive with their eyes, and hear with their ears, and understand with their heart, and turn for me to heal them.'

But blessed are your eyes, for they see, and your ears, for they hear. Truly, I say to you, many prophets and righteous men longed to see what you see, and did not see it, and to hear what you hear, and did not hear it.

§1b "Hear then the parable of the sower. When any one hears the word of the kingdom and does not understand it, the evil one comes and snatches away what is sown in his heart; this is what was sown along the path. As for what was sown on rocky ground, this is he who hears the word and immediately receives it with joy; yet he has no root in himself, but endures for a while, and when tribulation or persecution arises on account of the word, immediately he falls away. As for what was sown among thorns, this is he who hears the word, but the cares of the world and the delight in riches choke the word, and it proves unfruitful. As for what was sown on good soil, this is he who hears the word and understands it; he indeed bears fruit, and yields, in one case a hundredfold, in another sixty, and in another thirty."

§3a Another parable he put before them, saying, "The kingdom of heaven may be compared to a man who sowed good seed in his field; but while men were sleeping, his enemy came and sowed weeds among the wheat, and went away. So when the plants came up and bore grain, then the weeds appeared also. And the servants of the householder came and said to him, 'Sir, did you not sow good seed in your field? How then has it weeds?' He said to them, 'An enemy has done this.' The servants said to him, 'Then do you want us to go and gather them?' But he said, 'No; lest in gathering the weeds you root up the wheat along with them. Let both grow together until the harvest; and at harvest time I will tell the reapers, Gather the weeds first and bind them in bundles to be burned, but gather the wheat into my barn.'"

§4 Another parable he put before them, saying, "The kingdom of heaven is like a grain of mustard seed which a man took and sowed in his field; it is the smallest of all seeds, but when it has grown it is the greatest of shrubs and becomes a tree, so that the birds of the air come and make nests in its branches."

§5 He told them another parable. "The kingdom of heaven is like leaven which a woman took and hid in three measures of flour, till it was all leavened."

§3b All this Jesus said to the crowds in parables; indeed he said nothing to them without a parable. This was to fulfill what was spoken by the prophet:

"I will open my mouth in parables, I will utter what has been hidden since the foundation of the world."

Then he left the crowds and went into the house. And his disciples came to him, saying, "Explain to us the parable of the weeds of the field."

He answered, "He who sows the good seed is the Son of man; the field is the world, and the good seed means the sons of the kingdom; the weeds are the sons of the evil one, and the enemy who sowed them is the devil; the harvest is the close of the age, and the reapers are angels. Just as the weeds are gathered and burned with fire, so will it be at the close of the age. The Son of man will send his angels, and they will gather out of his kingdom all causes of sin and all evildoers, and throw them into the furnace of fire; there men will weep and gnash their teeth. Then the righteous will shine like the sun in the kingdom of their Father. He who has ears, let him hear.

§6 "The kingdom of heaven is like treasure hidden in a field, which a man found and covered up; then in his joy he goes and sells all that he has and buys that field.

§7 "Again, the kingdom of heaven is like a merchant in search of fine pearls, who, on finding one pearl of great value, went and sold all that he had and bought it.

§8 "Again, the kingdom of heaven is like a net which was thrown into the sea and gathered fish of every kind; when it was full, men drew it ashore and sat down and sorted the good into vessels but threw away the bad. So it will be at the close of the age. The angels will come out and separate the evil from the righteous, and throw them into the furnace of fire; there men will weep and gnash their teeth.

§9 "Have you understood all this?" They said to him, "Yes."

And he said to them, "Therefore every scribe who has been trained for the kingdom of heaven is like a householder who brings out of his treasure what is new and what is old."

§10 And when Jesus had finished these parables, he went away from there, and coming to his own country he taught them in their synagogue, so that they were astonished, and said, "Where did this man get this wisdom and these mighty works? Is not this the carpenter's son? Is not his mother called Mary? And are not his brothers James and Joseph and Simon and Judas? And are not all his sisters with us? Where then did this man get all this?" And they took offense at him. But Jesus said to them, "A prophet is not without honor except in his own country and in his own house." And he did not do many mighty works there, because of their unbelief.

The seven parables above are multiple views into a subject that Jesus understood deeply. He was delivering the parables so that those who received them could try to figure them out. If you could figure out what the parables meant, Jesus knew you had insight. The authors who wrote this story are from the insight school: They give insight to those who have it. They give understanding to those who understand.

They give knowledge to those who know. So when Jesus was asked, "Why do you speak to them in parables?" in section 2 above, he responded that there are those who have understanding, but have decided not to understand. There are those who have insight, but have decided not to test it. And there are those who have knowledge, but have decided not to increase it. From his perspective, which is consistent with the perspective of the writers through much of the Bible, this is the behavior that makes a person evil. Jesus had nothing to say to evil people and expects that they will lose the insight, knowledge, and understanding they have.

To solve this puzzle, we must again ask what this collection of parables, which is really a single riddle, is about.[45] The subject of each of the riddles is something Jesus calls the "kingdom of heaven." The parables explore the properties of this kingdom and our relationship to that kingdom.

As we look at the parables, we see that those in sections 1 and 3 above are longer and are told and then explained in detail. The parable in section 8 above is also long like the one in section 3, but it is not explained. The parable in section 8 is the last in the series, and it gives us the key we need to determine what this collection of parables is about. Angels are mentioned in this last story, and we've determined that these writers understood angels as relationships that both describe and control the world. This story is saying that there are relationships that operate on things in the world to separate them into those that are good and those that are bad. A world of gray becomes a world of light and dark. Each object in the world is known to be good or evil.

Our ignorance of the world is tremendous. Our understanding of the world is massively incomplete. So as we come to know more about the world, we have to determine whether what we learn is accurate. As we increase our understanding of the world, we have to determine whether our understanding accurately models the world. This story says that there are relationships that can be applied to our knowledge and understanding that can bring certainty to the uncertain. So in section 8, the kingdom of heaven is like the net filled with all kinds of fish. No one knows whether a particular fish is good

---

45 As I mentioned in Section 1.2, Jesus was all about understanding. The author repeats the word five times in Matthew 13. It's important. Jesus wants us to understand him, even though he is making one confusing statement after another. This behavior may not make a lot of sense to us, but it made sense to many of the intelligent people of this culture.

or bad—the net is full of fish uncertainty.

This good/bad mix of fish is the same as our world of gray mornings and gray evenings of Genesis 1. And the kingdom of heaven is about gathering lots of knowledge and understanding into one's mind, and keeping the good knowledge and understanding and discarding the bad. The angels are the relationships that allow us to determine what is bad and what is good. So what Jesus is talking about in this little riddle is *testability*, the idea that the predictions we make from our understanding must be tested to determine the faithfulness of that understanding. Unfaithful understanding is bad, and faithful understanding is good. Testability also involves testing what we observe of the world to be sure that the knowledge produced from those observations is accurate.

When a person has an insight, he does not know whether the understanding that results from that insight is good (correct) or evil (incorrect) until it is tested. The idea that understanding and knowledge must be put to the test is the concept being described here. The angels in this story are that concept. When the angels are done sorting the fish, there is no fish uncertainty. Every fish that's left is a good fish. That collection of fish is the kingdom of heaven which is, in fact, the Face of God, or UMkR. As we discussed earlier, God is composed of three things: the objects in the world, the relationships that control the world, and the UMkR that generates every relationship that describes and controls the world. This new world contains no uncertainty—it is deterministic. Everything is known by, and is under the complete control of, the UMkR.

Is this the world that we know? It most certainly is not. And these authors understood that. Their world contained a lot of uncertainty. The existence of that uncertainty told them the UMkR did not have complete control over the world. So they had to resolve that problem in their understanding of creation. They needed to find the reason why God did not have complete control over everything. The obvious reason was that they were wrong about God and that the UMkR was a model of something that could not exist. That conclusion was inescapable, but it was also unacceptable. Because God had to exist, and because it was clear that God did not exist where they were, they proposed that the UMkR would exist someday, or that it would exist in some parts of the world, even if it did not exist in others. For instance, the UMkR could exist in people like Enoch who were transformed and disappeared. It was clear, though, that God was out of reach and not in complete control, so they proposed that someday the UMkR would impose itself on all of creation.

What Jesus did here was to issue a prediction that the structure of the world must ultimately change—that God would one day exert control over everything that exists. This is not a new prediction. Earlier prophets also predicted that the world would change in a fundamental way. When that change happens, the UMkR will enforce its control over the world, and everything will become "deterministic." This changeover was known as the "Day of the Lord" and is supposed to be accompanied by the arrival of the "Ancient of Days," as mentioned in Daniel 7. The Ancient of Days would then establish his dominion over the earth.

Jesus is speaking in the same terms. The kingdom of heaven is God (as the UMkR) and the "end of the age" occurs when the UMkR imposes itself on the world we see around us (or any other region of the world, if the imposition does not occur everywhere). What Jesus is saying in the riddle of section 8 is that we can help our current world move toward this new world when we apply the concept of testability to what we know and understand of the world. When all of our models are tested and known to be good, they approach the single, all-encompassing model that perfectly models the world around us. That model would be the UMkR. We bring that perfect kingdom, that perfect understanding, that perfect angel into existence as we generate testably accurate prediction after testably accurate prediction. This is the message Jesus wanted to convey in this parable/riddle.

We see this story told again in section 3 above. The riddle in this section is divided into two parts. The first part is the riddle itself, and the second part is the explanation of the riddle. The riddle contains markers that relate it to the other riddles and goes on to add more detail to the overall story. The angels here are the same as those in section 8. The angels are, in fact, the fire. That fire represents the tests that show whether something is, in fact, evil. What was not known to be evil is now known to be evil. Once something that's wrong is known to be wrong, it can be avoided or discarded. For the writer of Matthew, to avoid doing evil destroys it as surely as fire destroys what is thrown into it. The thing that survives the fire is good. The symbology of the good surviving the testing of fire is often written about as gold that survives the purifying and separating fire of a furnace. Choosing the right and rejecting the wrong is known as judgment, and the fire symbolizes that judgment in many places in the Bible. The knowledge and understanding that survive the fire of the judgment are known to be right, and the purity of that righteousness is symbolized by the purity and intensity of the light emitted from the kingdom, the UMkR.

As we continue with the second half of section 3, the good seed represents the desire to remain alive, and the weed seeds represent the desire to experience pleasure. They are very often the same thing, but over time, one leads to life, and the other leads to death. When the authors state that they will reveal knowledge that "has been hidden since the foundation of the world," they are talking about the concepts, the philosophy, we're discussing here. What has been hidden?

- The concept of God that these authors use,

- the realization that the UMkR does not exist in any kind of world we are familiar with,

- the need for testable models of the world,

- the idea that the UMkR will someday impose itself on the world and eliminate ignorance and uncertainty, and

- the knowledge that seeking pleasure is not the same as seeking life.

Section 1 above provides an interesting variation of the story. The parable in that section is about whether and how well people understand that "the close of the age" is the coming of the kingdom of heaven (a world under the control of the UMkR) and whether they are willing to participate in making that world a reality. Helping to bring the kingdom of heaven into existence requires a dedication that few people possess. To make the new kingdom a reality imposes the absolute requirement that we only do what is good. We cannot perform a single evil act. That goodness requires that we must test everything we think we know to ensure that what we think is good is actually good. We have to test what we think we know to make sure that what we think is evil is actually evil.

When we are willing to live with uncertainty, we cannot bear fruit. In this riddle, the amount of fruit that each seed bears indicates how much that seed has helped advance the coming of the kingdom of heaven. Advancing the cause of the new kingdom is a pleasurable thing to those who accomplish it, and the fruit also symbolizes that pleasure. The path to the new kingdom is, in fact, the path to everlasting life. Those who are tricked by the pleasures to be had along the path are portrayed as those who don't ever understand the concept or the need for testability. Others tricked by those pleasures are portrayed as those who understand the concept of testability, but do not understand the need for it. Those people rely on the understanding of others. The authors say that they have no "root" in themselves. The final trickery is of those who actually understand but are tricked by the benefits of their great understanding into seeking pleasure for pleasure's sake instead of life for life's sake. Only those who seek life, instead of the pleasures of life, ultimately experience the pleasure of living forever in this kingdom. Those pleasures are told as people bearing fruit up to 100-fold.

At the end of the story of the Garden of Eden, God put in place a group of cherubs and a fiery, ever-turning sword to guard the way to the tree of life. The reason that the sword and the fire were used as symbols for the UMkR is made clear by this story: the sword represents testability, and the fire represents judgment. The sword tests and the fire destroys those things that do not pass those tests. Testability is the path to the tree of life.

The parable of the mustard seed in section 4 above symbolizes the concept of God that these writers held. The mustard seed is the UMkR or the Face of God, and it generates all of the relationships in the world (the branches of the tree). Those relationships energize the objects in the world (the birds that nest in the branches of the tree).

The parable of the leaven in section 5 above shows us another view of the concept of God that these writers held. The UMkR as the Face of God (the leaven) is the source of the spirit that energizes everything (the flour) in the world.

The parable of the treasure hidden in a field in section 6 above tells us that the world (the field) is a riddle produced by God (the treasure is covered up), and the solution to that riddle is something to be greatly prized. Those who know that there is a solution keep that knowledge hidden. The treasure is life everlasting, or the UMkR. The value of everlasting life is so great that anyone who has a shot at acquiring it would give up every pleasure in this life (money being stored pleasure) to acquire it.

The parable of the merchant seeking fine pearls in section 7 above tells us about those who are actively seeking everlasting life. When they find that the path to everlasting life must pass through the door of testability, they will also give up everything to follow the concept of testability to its logical conclusion.

In section 9 above, Jesus wanted to know if his disciples understood all the riddles he delivered. They assured him that they did, and he told them his teachings were different from those of the scribes. He said this to make the point that he was not rejecting the scribes' teaching, he was enlarging it. When Jesus taught the "close of the age" and the scribes taught "the Day of the Lord," they were talking about the same day. That day can occur for individuals like Enoch and Ezekiel, who did not die but were taken by God because they became completely righteous. Both Jesus and the scribes believed that the day would also occur for the world at large. The (slightly) new thing that Jesus taught was that the coming of the Day of the Lord was under our control. He decided to teach the masses so many more of them could practice testability, and the resulting increase in righteousness would cause that day to come more quickly for them individually and for the world at large.

In terms of the overall story, Jesus was teaching that we have control not only over the darkness-and-light mix of our minds, but also over the darkness-and-light mix of the world at large. When we are done transforming all that is unknown into the known, then there will be only good and evil and nothing in between. The final chapter of the book of Revelation shows this as the city of God filled with light and the outer darkness being kept at bay by the walls of that city. There will be no gray.

7.5 Existence

One of the more difficult things I had to conclude as I read through the Bible was that the God it described did not exist. It took longer, but I also concluded that the Bible writers were completely aware that the God they were describing did not exist. To be precise, it was the UMkR (the Face of God) that did not exist. The Bible writers understood, however, that it was useful to invent the UMkR. They knew that the UMkR was the result of a process of abstraction. In their everyday lives, they knew the world behaved in a more or less consistent manner. When they looked deeply into how things actually worked, they concluded that there were relationships between things, and those relationships somehow imposed themselves on objects in the world and controlled the behavior of those objects.

The next step they took was to explain how a world of relationships could work. When they realized that similar relationships were related by more abstract relationships, they made their first steps in the process of seeing the Face of God. They then turned this small abstraction into another, and another, and another. As they were ascending this pyramid of abstractions, they looked at where they were going, and that look was the inductive leap that helped them see a single relationship at the apex of that pyramid. That relationship was the UMkR, and it controlled everything. The result was a model of the world and a model of God. The Bible writers tell us about the creation of this model in Genesis 4:26. They tell us that men began to call upon the Lord by name. When they gave God a name, it referred to the model they had developed, and the creation of that model was the invention of God.

We create and deal with models all the time. A similar model to this model of God is the model we call a number. A number is a model of the world that represents the extraction of the idea of cardinality from objects in the world. For us, the number 5 does not control how many arms a starfish has, but it is clear that we have a mental model for the number. Does the number 5 exist? Does the Face of God exist? The number 5 certainly does exist as a model in my mind and in yours. It is a mental object that can exist within a mind. We know that. The Face of God can exist in the same way. If we project our mental models onto the real world and then behave as if those models exist independently in that world, we will also be forced to ask ourselves if the number 5 exists out there independent of you, independent of me, independent of human beings. That's a much harder question. The equivalent question is to ask, "Does the universe think?" Because if it does, then an independent mind exists that can hold the mental model we call "5." If a universal mind exists, then a mental model can have a separate existence.

The independent existence of the number 5 is a big question with huge consequences. It's as big as the question of whether the UMkR exists. From our perspective, the obvious answer is that neither exists in the external world separate from human beings. A much more limited question that addresses the same issue on a smaller scale is, "Do other animals think?" And the answer to that question, clearly, is that they do. It might be possible for another species of animal to have a mind that can hold the concept of cardinality, and the numbers for which they have models would have an existence that is independent of us. This would not be universal existence, but the mental model for the number 5 would exist independent of humanity.

The next, slightly more expansive question is whether the process of evolution has a "concept" of cardinality. That is, when we note that a starfish has 5 arms, we know that the cardinality of its arms is determined by its genetic makeup. But the only way that we could make the argument that the process of evolution has a model for the number 5 is if we could first make the argument that the evolutionary process contains or constitutes a mental process. It has been my opinion for some time that evolution is the process of life thinking about how to survive in its environment on earth. Conversely, I've also decided that the process we call "thinking" is the process of evolution sped up to operate over milliseconds on mental objects rather than operating over thousands of years on family groups. So I expect that someday we will find within our genetic programming actual models of the world.

The reason I think these models will be found is that the process that produced our minds is intrinsic to the operation of life on earth, so it should produce a "mind" or "minds" in our genetic material as well.[46] Our genetic material must be used to anticipate what will happen in the environment to help insure the survival of its host. That is something all living things (with genetic material) must do, and collections of living things must also anticipate what will happen. To ensure their survival, complex living things will not simply rely on reproduction and large numbers with passive recovery from events that threaten their species (even though that strategy *is* highly successful). In the case of humans, I think our minds developed into what they are because other humans became the greatest threat to human survival.[47] The need to anticipate the behavior of someone who is anticipating *your* behavior caused the human modeling system to greatly exceed that of all the other animals, so much so that it

---

46 I am being a bit inaccurate here. Our genetic material is an information store, so it is the "soul" of the energized mind that executes on the "processor" of the mass of living beings that seek to remain alive. Genetic material is not "alive," though we use that term to describe it. It "lives" as the instructions within it are executed on (also known as being "expressed within") the underlying living beings.

47 Though this would not have been the only reason. I think we became who we are because individual humans developed the ability to carry out threats with little or no consequence to themselves.

changed us into a different kind of animal.

This same thing is true for life in general. One species can be totally lethal to another. It is in the interest of those things that survive to anticipate and adapt to lethal threats. That adaptation would be greatly facilitated by the development of models of the other species. Once this kind of modeling exists, then the evolutionary threat/response arms race can occur on the level of models as well. Once these processes have advanced to the point that they become true mental processes, then mental objects like numbers can exist within the process we call evolution. Models that reside in the genetic material of living things allow models like the number 5 to have an existence that is completely independent of human beings.

If evolution is able to manipulate modeling systems as well as models, then it is conceivable that evolution could create a model of God just as these authors have done. (In fact, evolution *did* create the model of God that's in the Bible.) If evolution created a model of God independent of our model of God, that model would have an existence completely independent of human beings.

We know that humans evolved the ability to create not only models, but entire modeling systems. So we know that the process of evolution, entirely by itself, produced a creature that can create and manipulate modeling systems. Since the process of evolution on earth developed spontaneously, and we expect that evolution will occur wherever life develops, we know that the external world, entirely by itself, generated a system of models that manipulates systems of models. If modeling systems, and systems that manipulate modeling systems, can be generated spontaneously in our universe, that tells us something about our universe. It could be telling us something *fundamental* about the universe.

Is the universe itself a modeling system that, as a part of its nature, spontaneously and recursively generates other modeling systems? If it does behave in this way, we can think of the universe as a modeling system fractal.[48] That is, the universe would be a modeling system

---

48 A fractal is a collection of numbers in a set that result from the repeated application of an operation on that set. It can be viewed as a kind of "rough mixing." Fractal sets start with a few members, and after enough operations on the set, the set can have billions of members. If you operate on the set an infinite number times, the set will have an infinite number of members. Fractals are distinguished by the extraordinary complexity they can achieve, given that they start as such simple sets. The important thing to remember about them is that individual parts of the fractal are often strikingly similar to the whole set. This "self similar" property is characteristic of fractals. Fractals can be used to model

that generates other modeling systems that, in turn, generate other modeling systems. So the universe, as a modeling system, generated the modeling system that functions within evolution, and that modeling system generated us and our minds. And our minds continue the process of generating models and modeling systems. We're part of a fractal universe.

The fundamental rule at the very top of this universal modeling system would be the Face of God, and we see that face when we finally understand everything there is to understand. That fundamental relationship would be a model that would produce the models we know as archangels, cherubim, and ordinary angels that control objects in the world. It would be the UMkR of our universe. It would be the topmost model. It would produce, describe, and control everything in the universe for all time, past and future.

I say all of this not to take back what I've said about the existence of God, but to explain it. I am not going to claim that the natural numbers (1, 2, 3, ...) exist in the external world. I am not going to claim that the power of the natural numbers as a modeling system ultimately controls everything we do. The natural numbers can be used as a modeling system, and it is a powerful one. Is it powerful enough to model the entire universe? I don't know. Yet talking about its power comes as naturally to us as talking about the power of God. The modeling system that the UMkR represents is a different kind of modeling system but is a similar abstraction. And just like the natural numbers, we should not expect a modeling system like the UMkR to exist independently in our universe. We might ask what it might mean for the UMkR to exist outside of our universe, and there may be something to that, but we would have to admit that such a theory might not be testable, even in principle.[49] Because the UMkR behaves like a fractal that generates all of the complexity we see from an extraordinarily simple set of rules, the idea that everything is predicted/controlled by this simple set of rules can change the way we view the world. Predestination, for one, is implied. There would be no free will, and God would not throw dice.

This idea about how the universe works through a fractal God may have more to do with the way our minds work than with the way the

---

extraordinarily complex data.

49 Religious faith, in which a position is taken that cannot be justified even in principle, may have to be the basis for any position taken here.

world works. I think that what I am outlining here is what the Bible writers concluded about the world. I could have come to that conclusion because these ancient philosophers' minds worked the same as ours do, possibly in the same flawed way. So I must hold any conclusions about the structure of the universe at arm's length, because I am acutely aware of how much I don't know about the world (my internal grayness).

These ancient writers may have embraced the conclusions they reached because those conclusions were the best they could achieve. I'm trying not to impose my thoughts on these writers, since it is hard to know what they actually thought. And that is especially true because they took such great pains to conceal what they thought. So I must ask myself, "Did these ancient thinkers have an understanding of the world that is as deep as the one I've outlined here?" I think so. I have tried not to underestimate these writers, and I have tried not to overestimate them as well. But I do think their philosophy implies reasonable things about the universe, and these authors, I think, were aware of what their philosophy implied.

We've been talking a lot about numbers, so let me show you what they are and how they exist. Numbers are related to sets, and sets are fundamental to any discussion of this kind. A set is a collection of things that we think of as a unit for whatever task we are concerned with. We can collect anything into a set. Ultimately, a set is anything or any group of things on which we have decided to place our focus. The members of a set have properties, and at any particular moment we can ignore as many of the properties of any member of a set as we want. When we've ignored every other property, we are left with distinguishability (the ability to tell one member from another).

If we were to create a set that has members whose *only* property is distinguishability, and we insist that it makes no sense to say that two separate things are indistinguishable, then our set will contain two members. If we were to further insist that the only thing we cared about in that set was that we could tell one member from the other, that set would *define* the concept of "2." If we then took that set and decided to operate on it repeatedly as if it were a fractal set, we could increase the size of the "number set." The operation we would perform on the set would be to "add a distinguishable member." After the first operation, we would have a set that defines the number "3." Then the number "4" would follow, and so on. After an infinite number of additions of distinguishable members (an operation we cannot perform except in our minds), the resulting set would define the natural numbers (a set that exists only in our minds). What is so very interesting about this process is that the natural numbers can be viewed as a fractal set, and that fractals, though a fairly recent development in the field of mathematics, are an interesting way to view this fundamental mathematical object. (Fractal behavior is implied because a set that defines a particular number is so intensely similar to the set that defines the previous number, even though the two sets may have no members in common.)

The set that defines the natural numbers does not exist in the external world. It is a mental construction that we use to model the world.

These authors clearly understood the concept of number, even if they did not understand it as we do. It is much harder to say that they understood what a fractal is, even though they understood mixing. But let me show you how easy really important fractal-like sets are to construct without understanding fractals at all.

The concept of a fraction requires the use of two numbers. Let's start with the fraction 1/2. We can make a steady but everlasting descent to 0 by adding a 1 to the denominator. The result would be the sequence 1/3, 1/4, 1/5, 1/6, and so on. We could get this result by adding 0/1 to each succeeding fraction. The sequence would forever approach the number 0 without ever reaching it. Another simple thing we could do to our starting fraction would be to add 1 to the numerator and 1 to the denominator. The resulting sequence is 1/2, 2/3, 3/4, 4/5, 5/6, and so on. The sequence would forever approach the number 1 without ever reaching it. Both sequences are shown together in Figure 7.1.

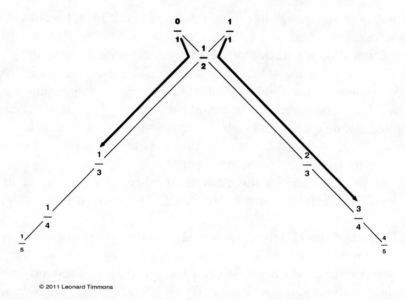

© 2011 Leonard Timmons

Figure 7.1: Two Sequences. Two sequences of related fractions can be generated by repeatedly adding 0/1 and 1/1 to the numerator and denominator of the fraction 1/2.

As you can see from the diagram, the improper fractions 0/1 and 1/1 can be used with the simple operation "add[50] root fraction to line fraction" to generate the next fraction in the sequence. At the limit on the 0/1 side, the sequence approaches 0. At the limit on the 1/1 side, the sequence approaches 1. You might notice that the fraction 1/2 can be generated by adding the numerators and denominators of the fractions 0/1 and 1/1.

If you notice that the root fraction is related to the line fraction by a change in direction, we can generate the set of all fractions by starting at a line fraction and changing the operation to generate the next fraction to "add line fraction to the first fraction up the line after changing direction." This slight generalization allows us to generate every fraction between the numbers 0 and 1. See Figure 7.2.

---

50 The numerators and denominators are added independently; they are not added like real fractions are.

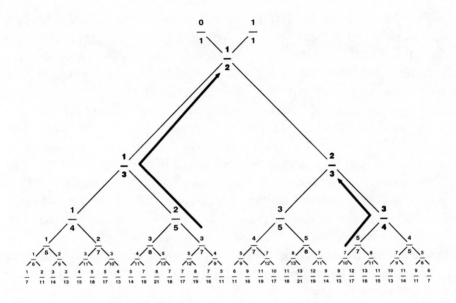

© 2011 Leonard Timmons

Figure 7.2: All Fractions. Using a simple rule and fractions for the number zero (0/1) and one (1/1), we can generate every fraction between the numbers 0 and 1.

The more generalized rule allows us to generate the fraction 2/5 by adding the numerators and denominators of 1/2 and 1/3. The next fraction along that line is generated by adding 2/5 and 1/2. The fraction 5/7 is generated by adding 3/4 and 2/3. The next fraction along that line is generated by adding 5/7 to 2/3.

This diagram shows an unconventional way to do a conventional thing—generate every fraction between 0 and 1. The great thing about this mechanism is that it gives us tremendous insight into the nature of all numbers. As you can see, each generation of fractions added to the set is twice the size of the previous generation. You might not notice that the golden ratio ($\sqrt{5}$ − 1)/2 and the square of the golden ratio trisect the diagram. If you follow the sequence 1/2, 2/3, 3/5, 5/8, and 8/13[51] by zigzagging forever downward, the limit would be

---

51 The Fibonacci sequence is generated by adding any two numbers in the following sequence to get the next number: 0,1,1,2,3,5,8,13,21,34... If we take the ratio of any two adjacent numbers (1/2, 2/3, 3/5, 5/8, ...), it approaches the golden ratio as the Fibonacci numbers get larger and larger. If we take the ratio of any two numbers separated by a single number (1/2, 1/3, 2/5, 3/8, 5/13, ...), then the ratio approaches the square of the golden ratio as the numbers get larger and larger.

the golden ratio. On the left, if you follow the sequence 1/2, 1/3, 2/5, 3/8, and 5/13 by zigzagging forever downward, the limit would be the square of the golden ratio. This result allows us to think of every irrational number (between 0 and 1) as a distinct and never-ending path through the diagram.

If you like playing with numbers, you will love this construction. I discovered it in the 1970s, I think. I have never seen it anywhere else, but I don't know if it's unique. My point is that this diagram is so incredibly simple that almost anyone could have come up with it. The Bible authors could have come up with it as well. I don't think they would have understood irrational numbers, but the golden ratio was known in ancient times. If you look more closely at the construction, you will find that the Fibonacci sequence is everywhere. The construction also allows us to rank fractions in terms of a concept of "complexity." The more complex the fraction, the further down the construction it occurs.

In terms of our current discussion, we can think of the zero (0/1) as Total Darkness and the one (1/1) as Total Light. Their interaction would create all of the complexity of the world. The irrational numbers would represent our path through the world. The mixing of light and darkness created the world as we know it. From our modern perspective, the world and each of us are moving down the diagram. From the perspective of these ancient writers, their world was moving up the diagram toward the separation of the universe into a world of Total Light and a world of Total Darkness.

7.5.1 Total Light and Total Darkness

For these authors, light and darkness are symbolic of something in the world. Light represents goodness in the world, and darkness represents evil. They refer to certain fundamental characteristics that all objects possess. We know that both observation and judgment are fundamental to their worldview, and we also know that the light and the darkness refer to their ability to observe the objects in the world. When we observe in the light, our observations produce good-quality data that can be relied upon. When we observe in the dark, our observations produce poor-quality data that is much less reliable. The good-quality data allows us to develop a more faithful understanding of the world, and the bad-quality data causes our understanding of the world to be much less faithful. So in a life-and-death situation where we have a real need to understand the way the world works, an unfaithful understanding can cost us our lives.

When our observations of the world are perfect and we produce perfect understanding from those observations, we are in Total Light. Being in Total Light means that our understanding is perfectly faithful, so the predictions we generate from that understanding give us perfect knowledge of what will happen. When we completely fail to observe and the lack of data causes us to fail to understand, we are in Total Darkness. Being in Total Darkness means that our lack of understanding prevents us from making any prediction about what will happen in the world.

Our observations of the world are never perfectly accurate and never perfectly inaccurate. The understanding we produce from those observations never produces predictions that are perfectly faithful. Our observations also never produce data that are in a form that resists all attempts to understand them. For these reasons, we must question whether Total Light or Total Darkness can exist in any world that we inhabit. And that question forces us to look at the concepts involved in Total Light and Total Darkness much more precisely. Because this philosophy is based on observation, and since modern science has a lot to say about what we can and cannot observe, we can examine in detail whether this philosophy has the ability to explain the way the world works.

An observation collects data into a set. We call a set of observed data a dataset. Each member of the set is a measurement, and each measurement is collected with a certain amount of uncertainty or error. When we observe, we collect data via our senses into an internal dataset. Suppose there were 81 members of a set of sensory measurements, and every measurement was exactly equal to 42. We could replace the dataset with a "relationship set" that contains an operator and a parameter that had the value 42, and the number of times to replicate that parameter's value. The operator would operate on the first parameter (the value 42) and use the second parameter to replicate that value 81 times to reproduce the original dataset. The operator would be the *relationship*. The parameters 42 and 81 are the knowledge we need to recreate the original dataset. This new relationship set of data replaces 81 pieces of data with three pieces of data, and that represents a massive compression of the amount of information we have to retain. This is what relationships do. They compress the data from the world into meaningful pieces of information. And *something has meaning if it can be used to make a prediction.*

When we recreate the original dataset from our relationship set, we are *predicting* that data. The accuracy with which we can recreate the original dataset tells us how faithful the relationship set is. If the original dataset is restored with little or no error, the relationship set is a faithful representation of that dataset (it has great faith). If the relationship set does not restore the original dataset within the margin of error we need, then that relationship set is not a faithful representation of the dataset (it has little faith). Faithful relationship sets, just like faithful relationships, are good things to have. Unfaithful relationship sets are bad. When our insight produces relationships in our minds, those relationships can be understood as internal relationship sets. When we require that a relationship set accurately reproduce the dataset from which it was abstracted, we are testing it. Untested relationship sets are gray, and we know that a relationship must be *tested* to determine whether it is faithful or unfaithful, good or evil, light or dark.

Total Darkness describes data in the world that has no meaning. If we could collect that data into a Total Darkness dataset, we could not find a relationship set that would describe that data. That is, there would be no way to compress the information contained in the Total Darkness dataset into a smaller set of data and an associated relationship. For this reason, the data contained within a Total Darkness dataset is completely nebulous, or totally meaningless. Given an arbitrary member of the dataset, you could find no operation you could perform on that member to produce another member of the set. This fact means there is nothing you can say about the value of that other member of the set. Your inability to make a statement about the value of the next member of the set implies that its value could not be measured (that is, that member could not be collected into the set). If the next member of the Total Darkness dataset was uncollectible, the original dataset member that we are supposed to be referencing must also have been uncollectible, and the Total Darkness dataset must be empty. So the Total Darkness dataset is the empty set, because the data in the set was never collected.

We can also look at the nonexistent data in the Total Darkness dataset as data that *was* collected by a process with a precision of measurement that was exactly zero. Or, said another way, the uncertainty associated with the measurement process was infinite.

And emptiness, by this definition, would be equivalent to a set that contains a measurement of zero precision or infinite uncertainty. Such a set cannot exist as a real-world dataset, because it is a contradiction in terms (no measurement has been made).[52] A Total Darkness dataset exists in our external world in the same way that the empty set exists, which implies that Total Darkness cannot exist, not in this universe.

The Total Light dataset is at the other extreme. Data in a Total Light dataset is completely and forever constant and is infinitely compressible. A Total Light dataset is an extrapolation from a dataset that contains extremely high-quality measurements that are highly compressible. Let's say that we have a dataset that contains 100,000,000 measurements. If all of the measurements were equal within our specified margin of error, then we could replace the dataset with a relationship set that contains an equality relationship, a single measurement, and the number of items in the original dataset. With these three pieces of information, we could reproduce the original 100,000,000 measurements within their margin of error.

We used the equality relationship in this case, but lots of relationships can compress to this extent. In particular, fractal relationships can compress an infinite amount of highly complex data into a very small starting dataset and a simple rule. That data would be tremendously meaningful. An example of what such a fractal relationship might be like is the relationship that generates the Mandelbrot set. The Mandelbrot set is a fractal set of incredible complexity generated by an extremely simple set of rules. Of course, the set of all fractions listed in order of complexity as shown in Figure 7.2 is also a complex arrangement of data generated by a very simple rule.

A Total Light dataset is at the absolute extreme of this kind of compression. It compresses to a single relationship that can be used to produce all the data in the Total Light dataset with zero error. This single relationship is actually a relationship set with *no* parameters. A relationship set for a Total Light dataset, with its single relationship, can reproduce every piece of data in the Total Light dataset with accuracy to its last decimal digit. The actual existence of a Total Light dataset would imply that one could achieve accuracy to the last digit in real-world measurements an infinite number of times. A Total

---

52 On the quantum level, the emptiness of space behaves, to some extent, like this kind of empty set.

Light dataset that describes (is a measurement of) everything in the universe does not exist. And with our current understanding of the universe, it *cannot* exist. The relationship set that would produce the Total Light dataset for our universe would be the UMkR.

Real datasets collected from the real world can be understood. They have meaning. But they also have an uncertainty that cannot be removed, even though it can be estimated. The uncertainty associated with real measurements is never zero and never infinite. So real-world datasets are necessarily gray. But some of those gray datasets are so light as to be almost Totally Light, and others are so dark as to be almost Totally Dark.

Complex living creatures like us have a need to know what is going to happen next. The predictions we make are critically important to our ability to maintain ourselves in a living state. The accuracy with which a living being can make predictions about its future (its ability to make those accurate predictions is its faith) can determine how well and how long it can maintain itself in a living state. Actions that maintain that living state are good actions, and the greater a being's ability to maintain itself in a living state, the more "good" that being is. Those of us who can predict what is coming with greater accuracy can use that knowledge to keep ourselves alive for an extremely long time. We often refer to people who have this kind of foreknowledge as being "in the light." Those of us who can't predict what is coming with any kind of accuracy or precision are "in the dark." It is in our best interest to collect high-quality data from the world. It is also in our best interest to generate faithful understanding from that data, so that our predictions can help keep us alive.

Our modern understanding of the world insists that uncertainty is embedded into the very nature of the universe. That view is compatible with the idea that Order and Disorder are mixed in the world. The modern view insists that Disorder will win unless there is some unknown process that will prevent it from winning. Our understanding is that Order and Disorder cannot spontaneously separate themselves. So this is where our view of the world would vary considerably from the view of these ancient authors. The modern view tells us that the universe is doomed to becoming more and more random over time until it becomes a disordered mess. The Second Law of Thermodynamics will make that happen. Everything must become gray, with no light and no darkness. The ancient view is that even though the total amount of Order and Disorder may remain the same, the Order will separate from the Disorder and refuse to mix with it.

We can reconcile the two views, however. The descent into grayness would not have to occur if a pure and infinite source of energy outside the universe (like Total Light, the UMkR) was constantly adding energy into the universe. A pure and infinite energy sink could also exist outside the universe (like Total Darkness), pulling energy out of the universe. Our world could then exist forever between the energy source and the energy sink. (Notice how the world would exist as a consequence of the flow of energy between an energy source and sink—just like living things do.) Another way to look at this would be to think of Total Light as an infinite source of Order, and its coupling to the universe injects Order. Total Darkness could be thought of as an infinite source of Disorder coupling Disorder into the universe. With these adjustments, the difference between the two views of the way the world works can be lessened. Of course, we have no evidence that there *is* an infinite source of energy, nor an infinite energy sink outside the universe. That doesn't mean that the concept would not be effective in modeling the world. In both views, the gray that is our world exists as a coupling between Total Light and Total Darkness. We can view that coupling as a Herculean battle for dominance at each and every point in space. These authors would say that the Total Light and Total Darkness are being separated (via testing) as the UMkR imposes itself on the world.

Another view of the way the world works is also possible. In that view, the world is not infested with disorder. That is, the UMkR has already imposed itself on the world and has total control of everything. That would mean that the universe is already deterministic, and our concept of "disorder as darkness leading to uncertainty" is flawed. The world might be so complex that it outstrips our ability to handle that complexity. That is certainly possible. We know that the universe is complex to an extreme that is hard to comprehend. We also know that in situations where we are overwhelmed by complexity, we have often *used the laws of probability to quantify our ignorance* of the process we are observing and trying to characterize. Modeling a highly complex but deterministic process as a random process can allow us to predict the gross behavior of that process. Those accurate predictions make our model useful, and that usefulness confirms that our model has value.

When we use probability to describe the behavior of the fundamental particles that make up the world, those probabilistic models could be describing an extraordinary complex but deterministic process. What we think of as a fundamental particle may actually be what we can observe of a much more complex process that we do not understand or cannot observe properly. When Albert Einstein rejected quantum mechanics because he could not come to terms with the use of probability to describe the fundamental behavior of our world, I think he moved toward this view of the world. Dr. Einstein famously said that God did not throw dice. In that statement, he was expressing a belief that our universe must be founded on deterministic rules, that those rules should apply at the very smallest scales of our world, and that those rules can be discovered. Underlying this belief is the idea that *everything* can, in fact, be understood. Everything.

But the idea that everything within the universe can be understood by something within that same universe is the ultimate issue. The ancients would have believed that some things could not be understood, things that were Totally Dark. And modern science has concluded you *cannot* know an object's position and velocity with ultimate accuracy. The ancient view is similar to the modern view. The ancients believed that Total Darkness and Total Light were both a part of the world. They believed that we could make ourselves a part of either. Their idea that they could know everything is the idea that they could make themselves some part of Total Light. For them, the knowledge of everything represents the complete eviction of darkness from their being.

The ancients would not have realized that a set that can model itself from within itself is a special kind of set. But we know that the group of natural numbers is that kind of set. We also know that the universe seems to be able to spontaneously create systems that can model the universe from within the universe (us, for example). So it has one of the critically important characteristics of the natural numbers. None of us have seen the entire universe or really know everything there is to know about it. But our ability to imagine what the universe should be like based on what we know allows us to imagine that it behaves like a system that is rich enough to allow something within that system to model the entire system. For us, that would mean it might be possible to understand absolutely everything about the universe.

The model of God that exists in the Bible is a projection of our understanding of ourselves onto the external world. That understanding, that self-model, did not always exist. When we understood ourselves well enough to create an extremely accurate model of ourselves from within ourselves, we became human. We later recognized the value of understanding as it gave us power over the world and gave us the ability to extend our lives. As our ability to model the world improved, some came to believe that it was possible to model, and therefore understand, *everything*. *The desire to understand everything* and the pleasure to be had as a result of fulfilling that desire became a tree of life to those who found it.

When the idea that one can understand everything is projected into the external world, it becomes the Face of God. Even though some might not have understood that their projection did not actually exist in the external world, some had to know that it was not there. These authors recognized that their ideal did not exist. But they also understood that the nonexistence of their ideal did not mean that they should not worship it. The fact that it was not there did not mean that they should not try to bring it into existence by actually trying to understand everything. It did not mean that they should not dedicate their lives to achieving what was impossible. It did not mean that they should not try to become a part of any thing that was moving toward that ideal.

The idea that we can understand the entire universe, that we can come to know everything, is another model of the world that we have extracted from the world. We don't know if such a thing is possible, but it was useful to invent the idea that everything is knowable, and it remains a useful invention. It *is* useful to try to know everything and to completely dispel the grayness of the unknown. To tell those who are infatuated with knowing everything that what they are trying to do is not possible is to give voice to an idea they have decided to ignore. They remain undeterred. Just as these writers did, they continue to advance toward a goal they have no reason to believe is actually there. To tell those who have determined that they must know everything, that they should set their eyes on more achievable goals is to insist that they destroy themselves. They have made the prediction that the knowledge of everything is the only goal that's ultimately worth reaching.

Knowledge seekers will tell you that the knowledge and understanding of everything is not only their single-minded goal;

they will tell you it is the goal of every single object in the universe. They will tell you that the pursuit of all knowledge leads to the pursuit of insight, and the pursuit of insight leads to the pursuit of understanding, and the pursuit of understanding leads to the pursuit of wisdom, and the pursuit of wisdom is the pursuit of life. This is what the stories of our textbook Bible were written to tell us. Are we willing to listen to what these knowledge seekers have to say?

7.6 Critical Knowledge

The discovery of a true solar calendar in Genesis tells us that the people who wrote Genesis (or edited it) were scholars of their time. This discovery also tells us that we must not underestimate what these ancient scholars knew, nor should we underestimate their understanding of the world around them. We might also conclude that the loss of this knowledge for such a long time may represent a "dark age"—if, in fact, the solution to this puzzle was actually lost.[dd] For instance, we know that the Essenes used a 364-day calendar, but they do not seem to have figured out the puzzle of the patriarchs' ages.[de] If the Essenes did understand the solution to this puzzle, and their system of education was one based on demonstrating insight, that system would require that they also hide what they knew within a riddle or puzzle.

Biblical critics[53] have discovered, through textual analysis, four separate, internally consistent documents that make up the first five books of the Bible. This discovery is known as the Documentary Hypothesis.[df] That analysis also seems to show that there were even more ancient source documents written by different people or groups of people at different times. One of those hypothesized source documents is called the "Book of Generations," whose title was taken from Genesis 5:1 (RSV) "This is the book of the generations of Adam." That book is supposed to be the source document that contained the genealogy of Adam through his son Seth. All of the ages of the patriarchs as recorded in Genesis 5:1–20, 30–32; 7:6; 9:28–29; and 11:10b–26, 32[dg] would have come through that source document. This view of the Bible has been extraordinarily successful in explaining the

---

53 Biblical critics are a kind of biblical scholar. They read the Bible as a literary text and analyze it with the understanding that it was written down by someone at some point in time. The analysis takes into account the writer and the environment in which the text was written, as well as what the author and subsequent editors were trying to communicate.

differences apparent in the text. If I had been as familiar with this analysis as most scholars are, I might never have considered Genesis 1 to 11 to be a unified whole. The Documentary Hypothesis is compelling.

I have two problems with the Documentary Hypothesis, however. First, no one was aware of the existence of a calendar hidden in the genealogy of Genesis 5 and the story of the Flood. I don't see how the Documentary Hypothesis can possibly explain the existence of this puzzle. As a part of that explanation, the Documentary Hypothesis would also have to explain the existence of the numbers in the Flood story that provided the intra-year divisions of the calendar, and it would have to explain the puzzle in Genesis 11 that confirms the calendar in Genesis 5. My analysis also shows that the genealogy of Cain in Genesis 4:17–26, is an integral part of Genesis 5. Since these verses are not a part of the proposed "Book of Generations" from which Genesis 5 is supposed to originate, the Book of Generations would have to be modified to include portions of the previous chapter, portions of the Flood story, and all of Genesis 11.

The second reservation I have is that given the complexity of the riddles and puzzles and the depth of thought shown by these authors, I've concluded that a person who could cut and paste together, essentially unmodified, independently written stories into this kind of unified whole could not possibly have existed.[dh, 54] On the contrary, it would seem much more reasonable that the portion of the Bible represented by Genesis 1 to 11 was edited as a whole each time it was revised, since it seems that the editors likely understood the story and the intent of the original writers. So the existence of the calendar and its confirmation argues emphatically for a single, precise composition, even if that composition was edited many times over many generations. So I think that the calendar puzzle in Genesis 5 falsifies the Documentary Hypothesis.

To create the intertwined voices in the text of the Bible that the proponents of the Documentary Hypothesis have so cleverly separated, the Bible writers could have intentionally written different verses in different voices. They may have composed the stories in this way to pay homage to an oral tradition in which multiple speakers

---

54 In *Who Wrote the Bible?*, Richard Friedman states that a number of the original sources are included without a single word being changed. In my opinion, the existence of the calendar makes that impossible. No one is that good.

delivered a single story. Over time, the material might have been modified to accommodate changes in language, for example, or to include the new insights of the next generation of wise men. When a new group of writers introduced new material, they could have used a voice from the original story for their new material, introduced a new voice, or restructured the entire story with a different set of voices. A proficient scribe would necessarily have had the ability to precisely mimic the voice of a speaker when modifying that speaker's words. I think that for a scribe, the ability to create new material in an established voice would have been an ordinary skill. I've done the same in my own life by writing a story in the voice of Kahlil Gibran's *The Prophet* years after scribing the entire book.[55] So creating the original story in multiple voices is a simple extension of an ordinary skill. When the Documentary Hypothesis recognizes these voices and concludes that they were necessarily different people, it is incorrect. It is a misreading of a very sophisticated work and an underestimation of its authors.[56]

I will agree with the biblical critics that the Bible writers did not believe in God as we commonly do (as I have detailed above), and the writers certainly did not believe as their followers in the general population did in their time. Within the community of critics, an idea exists that the Bible writers put together many of the stories in the Bible with full knowledge that those stories were untrue. The idea that the writers knew that their stories were untrue at the time they were written implies that the writers were lying to their readers. The story of the Creation was put on paper (or a similar medium) for the very first time at some point in history. The person or group who wrote it would certainly know that the story was not literally true, so writing the story implies that the writer or writers were being dishonest. My position here is that the stories were meant as teaching aids. The differences between our culture and theirs and our

---

55 When I was in high school, I had a manual typewriter. I also possessed a copy of Kahlil Gibran's *The Prophet*. I enjoyed it immensely. What I decided to do was to use it to learn to type better. So I copied the entire book using my manual typewriter. When I made a mistake on a page, I would discard the page and retype it from the beginning. I eventually completed the entire book without a single error (as far as I could tell). A few years after, I wrote an essay for a college English course in the style of *The Prophet*. My professor exclaimed that were I to enter my essay in a Gibran write-alike contest, I would win hands down. I had captured his voice and writing style by simply doing the job that a scribe does.

56 I think we've underestimated the ability of later writers to write in earlier styles, even to the point of using archaic dialects.

prejudices toward these writers have prevented us from understanding them and their intent. The entire idea that the writers intentionally told falsehoods in Genesis 1 to 11 for the spiritual benefit of their readers (a "pious fraud") is wrong, in my view.

In his book *Who Wrote the Bible?*, Richard Friedman makes clear that ancient Israel had two schools of thought about what our relationship with God should be. The first school believed in angels, quick wit, and direct intervention by God; the second believed in laws, plodding, and obedience to God. When we first began to talk about these issues in Chapter 3, I noted the difference between our system of education and the system of education I've proposed for this culture. We've all experienced the two ways to teach and to learn, and we can imagine how they can be at odds with one another. The insight school emphasizes the ability to leap to a solution after considering a matter. The rule-following school emphasizes the need to learn by prescription and by rote. The insight school moves us into the future and is based on a talent you either have or don't have. The rule-following school moves us through each day and tries to make effective use of what is already known. These two schools are not really at odds, they are complementary. In actual practice, one way of learning is often preferred over the other, and that preference is the source of the conflict. A balance can be struck between these two schools, though it is hard to know where that balance should be. People often lurch from one extreme to another, so societies are often at one of these extremes of belief.

An excellent example of the conflict between these two schools is given in the story of Jesus and Nicodemus. Jesus tells Nicodemus that he must be "born anew," and Nicodemus does not understand him. Jesus says, "Are you a teacher of Israel, and yet you do not understand this?"[di] Nicodemus was a member of the rule-following school, and Jesus was appalled that though he was responsible for teaching others, Nicodemus could not decode a simple riddle. The rule-following school was in ascendancy, and they had dispensed with the riddle-making of the insight school. Nicodemus would have been chosen to receive his education because of his position in society, not because of his innate ability to be insightful. Jesus, on the other hand, was a teacher with massive insight who could use riddles and puzzles with a dexterity that was extraordinarily uncommon.

Each of these schools can veer to its own extreme. The insight school can degenerate into teaching cryptic, mindless puzzles and riddles that require greater and greater ability to solve cryptic, mindless puzzles. Members of this school at this extreme would let their imaginations run wild. They would not require that their theories about how the world works be tested or testable. The rule-following school, at its extreme, can degenerate into the canonization and worship of what has been discovered by its insightful ancestors. Members of this school would come to believe that any new insight is an attempt to reopen the canon and desecrate their God-given laws. They would not require that their theories about the world be tested or testable.

I think these two schools and the conflict between them arose naturally at the very beginning of human education, and that the conflict continues to this day.

7.7 What We Know

We know that the Bible is the primary textbook of an ancient system of education. We also know that the Bible writers began that textbook with their philosophy. We know that their philosophy is based on reliability and testability. In particular, it is based on reliable and testable insights that result from reliable and testable observations. We know that we gain understanding from our insights and that those insights can help us to have additional insights about what we already understand. This acceleration of insight leads to a more abstract and more powerful understanding of the world. This process of abstraction can be thought of as a pyramid of relationships that describe and control the world. We know that the relationships that describe and control other relationships are meta-relationships, and at the top of this relationship pyramid is an Ultimate Meta-knowledge Relationship (UMkR) that corresponds to an Ultimate Understanding (UU) of the world.

We know this system of education was based on insight, and the Bible writers used riddles and puzzles to search for and to test for insight. The collection of their riddles and puzzles resulted in a document that formed the basis of their system of education: the Bible. The Bible is the textbook and the library produced by this system. Through these riddles and puzzles, the Bible writers communicated their philosophy that everlasting life is achievable via reliable and testable insights. And that philosophy tells us that God is the world around us, the relationships that describe and control the world, and the UMkR. It also tells us that the UMkR describes and controls every relationship, and that through those relationships it controls every physical thing in the world.

We also know that this is not the world we see and that the Bible writers knew this as well. So they looked toward a day when God would exert his control over all of creation and this ideal world would become our world. That day would be called the "Day of the Lord."

The energy behind the relationships that control the objects in the world is spirit, and that spirit originates with the UMkR and flows through all the relationships at each level of abstraction until objects in the world are energized. The UMkR is the Face of God. The meta-relationships and relationships together form the Body of God. Those relationships and meta-relationships, individually and as collections, constitute angels, cherubs, and other heavenly beings. The Body of God is also called the Son of God. The objects in the world are "God at rest," and they remain at rest until a relationship that controls them causes them to move.

We know that observation is a kind of measurement, and the data produced by an observation has an associated uncertainty. That uncertainty is not zero and not infinite. We also know that we can extract a relationship from that data and produce a smaller dataset. The relationship and the smaller dataset can be used to reconstitute the original data. We can repeatedly use this process of multilevel abstraction to produce relationships that represent deep abstractions of the original data. If we could produce a collection of data that represents all of the data in the universe, we could create simple abstractions of that data. Those abstractions are also data, and that data can be used to create more general abstractions. As we reach higher and higher levels of abstraction, the highest possible abstraction would be the UMkR. If we had a collection of data that represented a perfect observation of everything in the universe (a Total Light dataset), the UMkR would be the relationship that produces that set of data. It would produce a Total Light dataset for all time, both past and future. No parameters would be required to direct the behavior of the UMkR. It would "know" how to create the Total Light dataset for the universe.

We know that the desire to know everything is a tree of life for everyone who possesses it. The personification and projection of that desire into the external world is the God of the Bible. The fulfillment of that desire is the Face of God. Those who wrote the Bible understood that their God, as the desire to know everything, did exist, and the fulfillment of that desire, as the Face of God, did not exist. For them, God did exist and did not exist at the same time. That conflicting duality and mystery allowed the Bible writers to write on behalf of the God they worshiped without being wrong when they said that what they were writing was the "word of the Lord." Their unrelenting use of riddles and puzzles, and the difficulty of understanding their concept of God, has prevented us from understanding their writings.

We know that riddles and puzzles are an effective way to keep valuable knowledge hidden, yet available for others who have insight. The Bible writers encoded their view of human history, their mixture of science, history, philosophy, and theology, along with their calendar into the first riddles of their most important collection of riddles: the Bible.

We know that developing puzzles of this complexity and stories of this depth is not easy. It makes sense that the books of the Bible would be revised over time as different authors updated and improved the stories. For these authors, the Bible would have become their repository of reliable teachings, deep knowledge, and recognized wisdom. The authors of one era would enlarge the repository by adding their insights as new puzzles and riddles to be solved by their successors, and their successors would do the same. Cooperating in this way to gain a greater understanding of the world would have taken place over centuries. It would have worked in a way that is very similar to the way scientists work today.

We know that when Jesus traveled about announcing that everyone had to be "born again" and also calling himself the "Son of man," he was simply repeating the central idea of the story of the Garden of Eden. In that story, man became acutely aware that he was programmed to seek pleasure, and that seeking pleasure would ultimately result in his death. That realization changed him, and he became no longer man but the successor to man. He became the son of man. In the story of the Garden of Eden, Adam and Eve were born again.

We know that when Jesus spoke of having faith like that of a mustard seed, he did not mean that the mustard seed has little faith. The mustard seed has great faith, ultimate faith. That seed is symbolic of the UMkR, because it begins as the smallest of seeds and produces the largest of herbs. If you had faith like the UMkR, you would be able to move mountains, just like God can.

We know that living things, as we know them, always find themselves between an energy source and an energy sink. And we know that living things absorb energy from their energy sources to keep themselves alive. We also know that those living things try to remain energized as long as they possibly can. They would remain energized forever if they could. But they can't. Strategies to remain energized become tools in the living thing's quest for everlasting life. By their very nature, they seek a forever-reliable energy source. Also by that nature, they seek to avoid the forever-insatiable energy sink that guarantees death.

We know that these motivations are fundamental to life on earth. We are a part of that life, so we seek these same things. The absolute need to seek out the hoped-for-but-nonexistent, perfect energy source is not negated by the fact that it does not exist. The absolute need to absolutely avoid the unsatisfiably perfect energy sink is also not affected by the fact that it does not exist. We have personified the perfect energy source, the one that has the power to deliver to us the everlasting life that we seek, as God. And we have personified the perfect destroyer of energy as Satan.

We suspect that the universe we live in is like a living thing that exists between an energy source and an energy sink. Our universe might exist between a perfect energy source and a perfect energy sink, both of which are outside the universe itself. Our universe, like a living thing, continues to create things within itself that are very much like itself—it continually creates models of itself from within itself. That's what evolution is doing, and it's what our minds are doing. It is what our minds did when they created our model of God.

We know that to achieve the everlasting life we seek would require an infinite number of infinitely wise acts. But we also know that we have already confused seeking life with seeking pleasure. So our first step in the process of seeking life must be to give up the unbelievably beautiful lie that seeking pleasure is the same as seeking life. That is something we know.

We know that the calendar described by these genealogies is a purely solar calendar with overtures to lunar calendars, administrative calendars, and even an ecclesiastical calendar.[dj] We know that the Bible contains extensive calendar information, contrary to assertions made by some authors.[dk] We also know that this calendar is a practical and useful one based on a 364-day year that remains in sync with the seasons.

I must say that even though we think of our year as 52 weeks long, the idea of a practical 364-day calendar of 52 weeks exactly is the kind of thought that rushes into one's mind and leaves it just as quickly. Everyone knows that the calendar, as a problem, has already been solved. When it became clear to me in Figure 2.5 and Figure 2.6 that the puzzle authors wanted me to look at Enoch's age as 364 + 1 years long, I didn't have the slightest idea why. I did know that Enoch's age was clearly set to the number of days in a year, but setting the year to 364 days *exactly* did not make sense to me. I did not know that the Essenes had insisted on using a 364-day calendar, 500 to maybe 1,500 years after this calendar was invented. So by helping me find my way to the solution to their calendar puzzle, the Bible writers have achieved their goal of leading a willing initiate to the insights they wanted to pass on to their successors so many thousands of years ago.

And as they still went on and talked,
behold, a chariot of fire and horses of fire separated the two of them.
And Eli'jah went up by a whirlwind into heaven.
And Eli'sha saw it and he cried,
"My father, my father! the chariots of Israel and its horsemen!"
And he saw him no more.
2 Kings 2:11–12a (RSV)

Now when Eli'sha had fallen sick with
the illness of which he was to die,
Jo'ash king of Israel went down to him,
and wept before him, crying,
"My father, my father!
The chariots of Israel and its horsemen!"
2 Kings 13:14 (RSV)

Notes

aa Carol A. Hill, "Making Sense of the Numbers of Genesis," *Perspectives on Science and Christian Faith* 55, no. 4 (Dec. 2003): 239–251. http://www.asa3.org/asa/pscf/2003/PSCF12-03Hill.pdf (accessed December 16, 2010).

ab Andrew P. Kvasnica, "The Ages of the Antediluvian Patriarchs In Genesis 5" (Dallas: Biblical Studies Press, 1996), http://www.bible.org/page.php?page_id=3054, (accessed December 16, 2010).

ac Gerhard F. Hasel, "The Meaning of the Chronogenealogies of Genesis 5 and 11," *Origins* 7, no. 2 (1980): 53–70, http://www.grisda.org/origins/07053.htm (accessed December 16, 2010).

ad Richard Elliott Friedman, *Who Wrote the Bible?*, 2nd ed. (New York: Harper Collins, 1997), 15, 24. The Documentary Hypothesis.

ae I posted a number of articles to Usenet in 1995, 1996, and onward on these subjects. The articles, "Genesis/Eden/Garden/Solution," "Genesis/Eden/Garden/Solution/Full," and "Genesis/Noah/Flood/Solution," are directly on point.

af Genesis 4:17–5:32, 7:6, and 9:28 (JPS Tanakh).

ag Robert R. Wilson, "The Old Testament Genealogies in Recent Research," in *I Studied Inscriptions from before the Flood*, ed. Richard S. Hess and David Toshio Tsumura (Winona Lake, IN: Eisenbrauns, 1994), 200–223.

ah Duncan Steel, *Marking Time: The Epic Quest to Invent the Perfect Calendar* (New York: John Wiley and Sons, 2000), 381.

ai Ibid., 383.

aj Ibid., viii.

ak Ibid., 188–189.

al John 21:1–14 (RSV).

am Bruce K. Gardner, *The Genesis Calendar: The Synchronistic Tradition in Genesis 1–11* (Lanham, MD: University Press of America, 2001), 132–134.

an Leo G. Perdue, *Wisdom Literature: A Theological History*, 1st ed. (Louisville, KY: Westminster John Knox Press, 2007), 3.

ao Ibid., 32.

ap Ibid., 72.

aq Entry for "insight," *The American Heritage Dictionary of the English Language*, 4th ed. (New York: Houghton Mifflin Company, 2006).Copyright © 2009 by Houghton Mifflin Harcourt Publishing Company. Reproduced by permission from *The American Heritage Dictionary of the English Language*, Fourth Edition. See also http://dictionary.reference.com/browse/insight (accessed December 16, 2010).

ar John 1:1–3 (RSV).

as Genesis 1:1–3 (JPS Tanakh).

at John 1:5 (RSV).

au Hebrews 11:1 (RSV).

av Genesis 15:4–6 (RSV).

aw Revelation 4:8 (RSV).

ax Proverbs 1:1–19 (RSV).

ay Douglas R. Hofstadter, *I Am a Strange Loop* (New York: Basic Books, 2007), 266, 240–258. This idea is very well described in chapters 17 and 18.

az 2 Kings 2:1–18 (RSV).

ba Deuteronomy 18:18–22 (RSV).

bb Genesis 8:21 (JPS Tanakh).

bc Genesis 9:3 (JPS Tanakh).

bd Genesis 6:1 (JPS Tanakh).

be Genesis 9:7 (JPS Tanakh).

bf Genesis 3:18 (RSV).

bg John J. Parsons, "Nephilim. Hebrew for Christians Glossary Pages, Hebrew Glossary – N," in *Hebrew for Christians* (n.p.: Hebrew4Christians Ministries, 2007), http://www.hebrew4christians.com/Glossary/Hebrew_Glossary_-_N/hebrew_glossary_-_n.html (accessed December 16, 2010).

bh Geoffrey W. Bromiley, "Sons of God (OT)," in *The International Standard Bible Encyclopedia: Q–Z,* (Grand Rapids, MI: Wm. B. Eerdmans Publishing Co., 1988), 584.

bi India, "Who are the sons of God and the Nephilim?" in *Rational Christianity—Christian Apologetics*, http://www.rationalchristianity.net/nephilim.html (accessed December 16, 2010).

bj Genesis 6:4 (JPS Tanakh).

bk Geoffrey W. Bromiley, "Nephilim," in *International Standard Bible Encyclopedia: K–P* (Grand Rapids, MI: Wm. B. Eerdmans Publishing Co., 1986), 518–519.

bl Leon R. Kass, *The Beginning of Wisdom*, University of Chicago Press edition (Chicago: University of Chicago Press, 2006), 162. I make this reference to make it clear that others have come to the same conclusion I've reached here, though Dr. Kass does not seem to think that mankind was the first to eat meat.

bm The Committee on Scientific Evaluation of WIC Nutrition Risk Criteria of the Institute of Medicine, *WIC Nutrition Risk Criteria: A Scientific Assessment* (Washington, DC: The National Academies Press, 1996), 104–109.

bn Susan Pollock, *Ancient Mesopotamia: The Eden that Never Was* (Cambridge: Cambridge University Press, 1999), 40.

bo Barbara Ann Kipfer, "Longhouse/Long house," in *Encyclopedic Dictionary of*

Archaeology (New York: Kluwer Academic/Plenum Publishers, 2000), 317.

bp Ibid., "Hill fort," 236.

bq Habibollāh Āyatollāhi, *The Book of Iran: The History of Iranian Art*, trans. Shermin Haghshenās (Tehran, Iran: Center For International Cultural Studies, 2003), 19.

br Helena Hamerow, "Houses and Households: The Archeology of Buildings; The Longhouse," in *Early Medieval Settlements: The Archeology of Rural Communities in Northwest Europe 400–900*, (Oxford: Oxford University Press, 2002), 14.

bs Sidney Troy, *A History of Fortification From 3000BC to AD1700* (Barnsley, UK: Pen & Sword Military Classics, 2006), 7.

bt Peter Roger Stuart Moorey, *Ancient Mesopotamian Materials and Industries: The Archaeological Evidence* (Winona Lake, Indiana: Eisenbrauns, 1999), 333.

bu Ibid.

bv Graham Burrell and John Hare, *Review of HSE Building Ignition Criteria: HSL/2006/33* (Bootle, Merseyside, UK: Health and Safety Executive, 2006), 4.

bw John P. Hageman, Example 1: Retardant asphalt coating, United States, 4925494 (42 Susan Dr., Closter, NJ: 1990), http://www.freepatentsonline.com/4925494.html (accessed December 16, 2010).

bx Ahmet Ekmekyapar, Hürriyet Erşahan, and Sinan Yapıcı, "Nonisothermal Decomposition Kinetics of Trona," *Industrial & Engineering Chemistry Research* 35, no. 1 (1996), 258–262.

by John K. Warren, *Evaporites: Sediments, Resources and Hydrocarbons* (Berlin: Springer-Verlag, 2006), 79.

bz Helena Hamerow, op. cit., 14.

ca M.G. Easton, *Illustrated Bible Dictionary*, 2nd ed. (New York: Thomas Nelson, 1897), 51.

cb Habibollāh Āyatollāhi, op. cit., 20.

cc Genesis 6:21 (RSV).

cd Daniel E. Fleming, *Time at Emar: The Cultic Calendar and the Rituals from the Diviner's Archive* (Winona Lake, IN: Eisenbrauns, 2000), 133.

ce Genesis 5:29 (JPS Tanakh).

cf Genesis 2:4 (JPS Tanakh), emphasis added.

cg Leo G. Perdue, op. cit., 46.

ch Richard Elliott Friedman, op. cit., 22, where the repeated stories are called "doublets."

ci Bruce K. Gardner, op. cit., 54.

cj Carol A. Hill, "The Garden of Eden: A Modern Landscape," *Perspectives on*

Science and Christian Faith 52 (March 2000): 31–46, http://www.asa3.org/ASA/PSCF/2000/PSCF3-00Hill.html (accessed December 16, 2010).

ck John J. Parsons, "Eden. Hebrew for Christians Glossary Pages, Hebrew Glossary – E," in *Hebrew for Christians* (n.p.: Hebrew4Christians Ministries, 2007), http://www.hebrew4christians.com/Glossary/Hebrew_Glossary_-_E/hebrew_glossary_-_e.html (accessed December 16, 2010).

cl Genesis 2:9 (JPS Tanakh).

cm Ibid.

cn Genesis 2:16–17 (JPS Tanakh).

co Genesis 3:2 (JPS Tanakh).

cp Genesis 3:15 (RSV).

cq Genesis 4:1 (JPS Tanakh, Footnote b).

cr Genesis 4:6b–7 (JPS Tanakh).

cs Genesis 9:18–27 (JPS Tanakh).

ct Genesis 11:1–9 (JPS Tanakh).

cu Genesis 11:10–26, 32 (JPS Tanakh).

cv Isaiah 64:7 (JPS Tanakh).

cw Isaiah 45:9b (JPS Tanakh).

cx Douglas R. Hofstadter, op. cit., 207.

cy Ibid., Chapter 9, pp. 104, 113.

cz Paul-Alain Beaulieu, "The Social and Intellectual Setting of Babylonian Wisdom Literature," in Richard J. Clifford, ed., *Wisdom Literature in Mesopotamia and Israel* (Atlanta: Society of Biblical Literature, 2007), 5.

da Revelation 13 (RSV), annotations and emphasis added.

db Revelation 13:18 (ASV).

dc Douglas R. Hofstadter, op. cit., Chapter 5, pp. 65–71.

dd Bruce K. Gardner, op. cit., 8.

de Ibid., 159, 176.

df Richard Elliott Friedman, op. cit., 60.

dg Ibid., 256.

dh Ibid., 228.

di John 3:10 (RSV).

dj James C. VanderKam, *Calendars in the Dead Sea Scrolls: Measuring Time* (London: Routledge, 1998), 31.

dk Ibid., 8.

Index

N

nahcolite 103
naked 132
natron 103
Nephilim 98, 120, 136
Nephilim, New 99, 120, 174
Nicodemus 217
Nimrod 145
Noah 30
Noah's Flood
    See: Flood story
number set, the 202
numbers
    1,056 45
    100 141
    105 24, 25, 35, 38, 45
        as half of 210 38
        as symbolic of 777 38
    11 113
    112 110
    120 98
    15 104
    150 108
    162 as the "fix-up point" 45
    168 152
    182 110, 152
    183 110
    187 21, 162
        secret of 44
    21 as 7 + 7 + 7 26
    210 25, 38
        symbolic of 777 38
    252 21, 23
        as 3×84 years 23
        as 364 − (2×56) 23
        as 65 + 187 21
    30 103, 109
    300 104
    33 93
    350 152
    360 157
    364 104, 162, 221
    365 10, 93, 104, 162
    369 104, 155, 162
        composition 44
    40 108
    416 23, 26, 28, 31, 38, 39
    50 103
    500 45, 141

    composition 45
    52 39
    530 154
    556 45
    56 14, 23, 38, 110
        as (7×7) + 7 17
        as 65 21
        as symbolic of 777 23
    600 141
    65 21, 35, 42, 110
    666 176
    73 109, 110
    730 104
    77 49, 110, 112
    777 10, 13, 18, 38
    800 21, 31, 35
        as a time, two times, and half a
            time 38
        as symbolic of 777 38
    84 23, 152
        as 77 + 7 29
        as 840 29, 35
        as symbolic of 777 23
    91 27, 152, 158
        as 910 29, 35
    existence 202
    existence of 198
    zero 49
    "nice" numbers 9, 18, 19, 35, 38, 45
    "odd" numbers 9
    "round" numbers 9, 38, 45
Numbers Game 22
numeric symbolism 26, 38
    369 158
    416 and 777 38
    800 38
    and the Divisible Year 38
numerical proverbs 51
numerology 47, 49
nutrition, intense 97

O

observation 90, 206, 218
omnivorism 106
Order 210
original man 181